THE CITY

THE BASICS

The City: The Basics provides a brief yet compelling overview of the study of cities and city life. The book draws on a range of perspectives – economic, political, cultural, and environmental aspects are all considered – to provide a broad survey of the evolution of cities in the rich Global North and the poorer Global South. Topics covered in the book include:

- a brief history of cities from ancient times to the post-modern present
- the differences between "global cities" in the North and "megacities" in the South
- the environmental impact of urban life
- urban economics, urban politics, urban culture, and urban planning.

Featuring suggestions for further reading, a glossary, and several international case studies, this is the ideal starting point for those interested in any aspect of cities or urban studies.

Kevin Archer is Chair of the Department of Geography at Central Washington University. He teaches graduate and undergraduate courses on urban issues, globalization, and the production of nature.

The Basics

ACTING
BELLA MERLIN

ANTHROPOLOGY
PETER METCALF

ARCHAEOLOGY (SECOND EDITION)
CLIVE GAMBLE

ART HISTORY
GRANT POOKE AND DIANA NEWALL

ARTIFICIAL INTELLIGENCE
KEVIN WARWICK

THE BIBLE
JOHN BARTON

BUDDHISM
CATHY CANTWELL

CONTEMPORARY LITERATURE
SUMAN GUPTA

CRIMINAL LAW
JONATHAN HERRING

CRIMINOLOGY (SECOND EDITION)
SANDRA WALKLATE

DANCE STUDIES
JO BUTTERWORTH

EASTERN PHILOSOPHY
VICTORIA S. HARRISON

ECONOMICS (SECOND EDITION)
TONY CLEAVER

EDUCATION
KAY WOOD

EUROPEAN UNION (SECOND EDITION)
ALEX WARLEIGH-LACK

EVOLUTION
SHERRIE LYONS

FILM STUDIES
AMY VILLAREJO

FINANCE (SECOND EDITION)
ERIK BANKS

HUMAN GENETICS
RICKI LEWIS

HUMAN GEOGRAPHY
ANDREW JONES

INTERNATIONAL RELATIONS
PETER SUTCH AND JUANITA ELIAS

ISLAM (SECOND EDITION)
COLIN TURNER

JOURNALISM STUDIES
MARTIN CONBOY

JUDAISM
JACOB NEUSNER

LANGUAGE (SECOND EDITION)
R. L. TRASK

LAW
GARY SLAPPER AND DAVID KELLY

LITERARY THEORY (SECOND EDITION)
HANS BERTENS

LOGIC
J. C. BEALL

MANAGEMENT
MORGEN WITZEL

MARKETING (SECOND EDITION)
KARL MOORE AND NIKETH PAREEK

MEDIA STUDIES
JULIAN MCDOUGALL

THE OLYMPICS
ANDY MIAH AND BEATRIZ GARCIA

PHILOSOPHY (FIFTH EDITION)
NIGEL WARBURTON

PHYSICAL GEOGRAPHY
JOSEPH HOLDEN

POETRY (SECOND EDITION)
JEFFREY WAINWRIGHT

POLITICS (FOURTH EDITION)
STEPHEN TANSEY AND NIGEL JACKSON

THE QUR'AN
MASSIMO CAMPANINI

RACE AND ETHNICITY
PETER KIVISTO AND PAUL R. CROLL

RELIGION (SECOND EDITION)
MALORY NYE

RELIGION AND SCIENCE
PHILIP CLAYTON

RESEARCH METHODS
NICHOLAS WALLIMAN

ROMAN CATHOLICISM
MICHAEL WALSH

SEMIOTICS (SECOND EDITION)
DANIEL CHANDLER

SHAKESPEARE (THIRD EDITION)
SEAN MCEVOY

SOCIAL WORK
MARK DOEL

SOCIOLOGY
KEN PLUMMER

SPECIAL EDUCATIONAL NEEDS
JANICE WEARMOUTH

TELEVISION STUDIES
TOBY MILLER

TERRORISM
JAMES LUTZ AND BRENDA LUTZ

THEATRE STUDIES
ROBERT LEACH

WORLD HISTORY
PETER N. STEARNS

THE CITY

THE BASICS

kevin archer

LONDON AND NEW YORK

First published 2013
by Routledge
2 Park Square, Milton Park, Abingdon, Oxon OX14 4RN

Simultaneously published in the USA and Canada by Routledge
711 Third Avenue, New York, NY 10017

Routledge is an imprint of the Taylor & Francis Group, an informa business

British Library Cataloguing in Publication Data
A catalogue record for this book is available from the British Library

Library of Congress Cataloging in Publication Data
Archer, Kevin, 1957-
The city : the basics / Kevin Archer.
p. cm. – (The basics)
1. Cities and towns. 2. Sociology, Urban. 3. City life. I. Title.
HT151.A724 2013
307.76–dc23
2012015729

ISBN: 978-0-415-67080-7 (hbk)
ISBN: 978-0-415-67079-1 (pbk)
ISBN: 978-0-203-08466-3 (ebk)

Typeset in Bembo and Scala Sans
by Taylor & Francis Books

Printed and bound in Great Britain by
TJ International Ltd, Padstow, Cornwall

CONTENTS

1

CITIES AND CITY LIFE

For the first time in human history, most people now live in cities. This urbanization of the world's population is predicted to continue, even intensify, in the years ahead. According to one estimate, by 2020, global population will reach 8.1 billion and only 37 percent of this total will live in rural areas. The pace of city growth is especially rapid in poorer countries of the **Global South**, particularly in Asia and Africa, where the urban population is expected to double between 2000 and 2030. Indeed, a recent report by the United Nations estimates that by 2030, cities of the Global South will account for as much as 81 percent of the world's urban population. It is clear, then, that the urban condition, across the planet, is fast becoming the human condition.

In addition to describing the historical and contemporary processes behind these trends in global urbanization, this book is an attempt to determine, in broad overview, what this latter assertion means. In what ways does life in the city differ from that in rural areas? If it does, what are the differences and in what ways do these differences matter for the human condition, now and in the future? Some have argued, for example, that life in cities is so significantly distinct that it represents a whole new form of social relations demanding, in turn, new ways of thinking about one's neighbors, one's self, and one's place in a greater community. If this is the case, then it behooves those concerned about the global human condition to figure out just what the implications may be of such rapid urbanization on a planetary scale.

To study "the city" in this way it is necessary to take an interdisciplinary, holistic approach. There is nothing out there in reality that is the city "economy" that is not already related to city "politics," city "culture," even to the city's natural "environment." Inner-city poverty in the United States, for example, has as much to do with policy-making or a culture of insecurity as it does with job creation and markets. Similarly, a world of hyper-wealthy cities and hyper-poor cities did not just appear as a result of some natural economic evolution but, rather, as a historical legacy of political oppression and economic exploitation on a global scale. While the following chapters of this book focus specifically on one or another of these aspects of the city, this essential interrelationship among them needs to be kept in mind if one is to come to a comprehensive understanding of the city. At the same time, one needs to avoid reifying the "city" as if it were an active agent itself. The "city" does not "do" anything or act in certain ways but, rather, city people do, in various ways with various results.

CITY AND COUNTRY

The initial thing to notice about cities is that they consist of relatively large and densely packed groups of people crowded into small pieces of territory. This is at once a fundamental change from rural life where small groups of people sparsely occupy rather large territories. From this quite obvious beginning arise several less obvious characteristics of city life. First, this large group of city-people somehow must find a way to feed itself. The limited territorial extent of the city precludes the possibility of feeding all from within. Second, the fact that we are talking about a relatively large group of people means that the population of the city cannot consist solely of extended family members or a few well-known neighbors. Rather, it is a population of relative strangers, difficult, if not impossible, to get to know on anything more than a limited basis. In this social context, it is difficult to know who to trust or to whom to delegate decision-making authority for the whole. Third, this dense population of relative strangers necessarily includes individuals from all classes, rich and poor, coming from different villages, near and far, with different, and sometimes very different, customs. Because of close proximity in the city, it is difficult, if not impossible, to avoid interacting with such social and cultural strangers. And such interaction necessarily produces reflection on, even a questioning of, one's own social station and what appropriate cultural norms and customs might be.

Already, this overly brief account sets one on the path to thinking about some distinguishing characteristics of city life. In the chapters that follow, these characteristics will be discussed in more depth in terms of the specific economic, political, and cultural traits of cities. Before turning to these, however, it is necessary to highlight one further thing with regard to cities of increasing importance in the world today. This fourth aspect concerns the relation of cities to the natural environment. City populations necessarily build dense configurations of material and relatively permanent living, working, and symbolic monuments. Such monuments replace or otherwise extremely modify the nature of the place. Similarly, large, tightly packed human populations on small territories not only need food to survive, but also water for drinking, cleansing, producing goods, and flushing waste. Occupying such a small territory, city populations also need someplace, and increasingly someplace else, to dispose of their material wastes. Cities thus have a significant ecological impact which cannot be ignored and which some argue makes them inherently unsustainable as modes of human life. That this may be the case will also be explored in this book. Here, suffice it to say that if cities are, indeed, environmentally unsustainable, then the fact that the planet is so rapidly urbanizing should be cause for great concern for everyone. Again, this small book is intended as a broad overview of just what might be at stake in this respect.

THE GLOBAL SPREAD OF CITIES

Since the **Agricultural Revolution**, when much of humankind stopped being predominantly hunters and gatherers, there have always been cities in the world. However, since the Western **Renaissance** (ca. 1250), the nature of city development changed and, indeed, cities began to spring up and develop on a regular basis first throughout Europe and then in the non-Western world as a result of

European exploration and eventual domination. Unlike earlier cities which emerged and grew, if they did, essentially on the basis of religious or political reasons, the emergence and growth of cities after the Renaissance became more a function of changes taking place in the economic life of countries. As a result of Western imperialism after 1492, a world system of economic development centered on Europe emerged in which cities around the world were either developed or reoriented in order to facilitate the functioning of the system as a whole.

Contemporary city life both in the Global South and the **Global North** reflects this historical spread of Western economic domination. To understand today's cities it is thus necessary to look more closely and critically at this history, as will be done throughout the chapters of this book. Indeed, some argue that the economies of cities in the Global South remain in a **neocolonial** relationship in the still Western-dominated world-system, which continues to hinder their development even after formal imperial relations have ceased to exist. This explains, it is argued, why post-colonial cities have not and, in fact, will never achieve what has been called global city status in the world economy. Cities in the Global South, according to this view, will always remain economically dependent upon the Global North and such truncated development explains why ever-larger **megacities** in poorer countries is not an indication of economic success as much as failure. In fact, such megacities are really an example of **over-urbanization** or city growth that is out of all possible control or regulation, whether in social or environmental terms.

Others argue that this is not entirely the case. Particularly with today's increasingly rapid **globalization** processes, many cities in the Global South are undergoing economic transformation. More and more investment from richer countries is flowing into countries such as Brazil, India, China, South Africa and others as manufacturing firms increasingly **off-shore** important parts of their production process. Even some professional services, such as accounting, engineering, and health, are now increasingly off-shored as a result of global competition and this can only bode well for the economic future of cities in poorer countries. If one looks closely enough at most post-colonial cities one, in fact, can see rapidly growing business districts with the same type of office buildings, hotel-convention centers, restaurant and entertainment facilities, and shopping malls, all found in richer cities of the Global North. The city people who make use of such facilities also appear to have much in common with their counterparts in richer countries. Perhaps, then, these similarities in city life across the planet portend the development of a city-based, truly global culture?

EMERGING GLOBAL DOWNTOWNS ON A 'PLANET OF CITY SLUMS'?

But still others argue that if this is a global culture in the making, it will not be a culture that the majority of the world's population will take part in, particularly in poorer countries. It may be that central cities around the world are becoming increasingly alike in their built and social environments. But even in the Global North there are significant areas of poverty and deprivation closely adjacent to the gleaming downtowns of built and social similarity. If even in so-called global cities such as London, Tokyo, and New York there are slum-like conditions for a good portion of the city population, it is even more the case in very rapidly growing cities such as Rio de Janeiro, Mumbai, Guangzhou, and Johannesburg. In many respects, what appears to be happening is the development of global downtowns more connected economically, culturally, and even politically with each other than with city dwellers in much closer spatial proximity.

Indeed, some have gone so far as to argue that the recent very rapid urbanization in the Global South is leading to a veritable "planet of city slums" where most city people barely eke out a decent subsistence in ever-overgrowing megacities. If this is the case, then such cities face increasing and quite daunting problems trying to provide adequate housing, health and emergency services, water and waste infrastructure, and security for ever-more people in no position to help pay for any of it. Given that the most rapid urbanization is taking place now in the Global South, as noted above, then this does not bode well for the future of the planet either in social or environmental terms.

Finally, if, as many now argue, the nation-state is losing power as a result of globalizing economic, political, and cultural processes, then what is actually happening in cities becomes even more important. Nation-states, for example, are finding it more and more difficult to control economic development within their borders as finances fly across the globe at digital speed. In turn, new supranational political-economic organizations such as the World Bank, the International Monetary Fund, the World Trade Organization, and the European Union now make decisions that nation-states must adhere to in order to remain viable members of the global political-economic system. Finally, digitally enhanced media are increasingly global in scope and, as a result, the dominant Western cultural influences that they project are, too.

Of importance, to the extent that nation-state power may, in fact, be declining, the role of cities as major engines of formerly national economies becomes that much more important on the global stage. It just may be that the global political economy is evolving in such a way now that cities will become, on their own, major players. It just may be, in other words, that current trends are creating a global system controlled largely by **city-states** not unlike that which existed before the rise of nation-states. If this is the case, then a better understanding of city life, in general, as well as of contemporary processes of urbanization in both the Global North and South, is absolutely necessary in order to make sense of today's globalizing world.

A WORD ABOUT FOCUS AND TERMINOLOGY

Given the intent of this book as a basic and broad survey of "the city" some hard choices had to be made in terms of focus and presentation. There has been a wide variety of city developments both historically and currently in what is now a global urbanization process. With more space allowed, of course, many more of these important developments could be treated in this text. Since this was not possible, I have focused on what I consider to be the most important trends, both historical and contemporary, that have made, and continue to make, for city development on a global scale. While many cities in many areas of the world have existed for a very long time, I put most emphasis on the processes that have made for what I call the Western[1] city over time, as well as the impact upon non-Western city life of Western city-based cultural intrusion via imperialism and colonialism after 1492. The belief is that, after 1492, city developments in the non-West essentially have involved a reckoning with more powerful economic, political, cultural, and even environmental forces intruding from this external source, even after the end of formal imperialism.

This does not mean that there are no remainders of earlier, and continually diverse, modes of city existence and development in the world today. Quite the contrary, Westernization has always been an uneven process both at the height of formal imperialism and within today's process of globalization. But to treat this unevenness properly would necessitate many more chapters than I have been allowed, so I am simply signaling this weakness in the narrative that follows at the outset. Similarly, there is some emphasis on the city experience of the United States in the chapters which deal with contemporary city life. This, again, was a conscious choice. City development processes in the United States exhibit the clearest examples of what I have called Western city life since they emerged and have evolved with little resistance, or remainder, from indigenous cultural processes. Moreover, that this is arguably the case renders the understanding of U.S. city processes even more important because they just may be the harbinger for city development across the planet as a result of the now global spread of neoliberal development policies.

Finally, the bounding of historical epochs is always a contentious exercise. In this book, the notion of "circa" (ca.) – that is, "around" that time or those times – is put to what some might consider overly liberal use. A more generous reading, however, would make clear that precise dating is less important than a consideration of the historical substance that renders such epochs relatively distinct. That is the position taken in this book via a rather unorthodox labeling and historical consideration of the following four phases of city development: Ancient/Classical (to ca. 1250), Mercantile/Colonial (ca. 1250 to ca. 1750), Industrial/Imperial (ca. 1750 to ca. 1970), and Post-Industrial/Post-Colonial (ca. 1970 to Present), with the last epoch more variable, of course, in terms of when former imperialized countries of the Global South actually attained formal independence. There is nothing sacrosanct about this phasing of city history; but it does underscore particular and significant changes that occurred in cities in certain periods of time that demand consideration. As for dating itself, to save repetition, the use of CE for Common Era and BCE for Before Common Era should be assumed for all dates that are noted in this book.

Box 1 Shifting city populations over space and time

Cities arose where humans were able to produce an agricultural surplus over and above day-to-day subsistence needs. Not surprisingly, it was in the fertile riverine environments of North Africa, the Middle East, Asia, and Mesoamerica where larger-scale, systematic agricultural production first emerged that cities themselves first appeared. These were areas that have come to be known as the cradles of human civilization for this very reason. With increased numbers of city people came the ever-deepening of human cultural achievement. In short, it was in precisely these areas of the world that the greatest civilizations of the ancient age emerged and prospered.

This is important to emphasize, particularly given that the main focus of this book is on the city experience of the West. Great cities came into existence in many regions of the world and have existed for a much longer period of time than those of the West treated so generously in the main text. And this is certainly reflected in the estimates of city populations over time (see Tables 1 to 4). Although the very rough estimates for Ancient/Classical, and even Mercantile/ Colonial, cities should be considered with no little caution, the overall regional pattern of city development and civilizational growth over a long history is clear enough. However, what is also clear is that this pattern has changed in significant ways, underscoring the main theme of this book. Many of the greatest cities and civilizations of the Ancient/Classical world were located in non-Western regions of the world to be sure. But the populations of Western cities began to grow rapidly and continually in the Mercantile/Colonial period and came to global domination in the Industrial/Imperial period. Furthermore, this domination continues until the effects of the Medical Revolution begin to be felt in the increasingly post-colonial regions of the world after 1950.

Indeed, this last point underscores something else emphasized in the main text and well reflected in the data on city population. Whereas in history, cities grew to the extent that available surpluses grew, whether procured in their immediate hinterland or over a more expansive empire, this link between surplus and population growth has not been forged in many of the largest and most rapidly growing cities of today. The population of these new megacities continues to grow, but the available surplus to support such growing populations has not been as forthcoming. As described throughout this book, such a situation certainly does not bode well for the future for many (and perhaps the majority of) city dwellers today.

Table 1

	200 BCE	*Estimated population*
1	Xi'an, China	400,000
2	Patna, India	350,000
3	Alexandria, Egypt	300,000
4	Seleucia, Iraq	200,000
5	Carthage, Tunisia	150,000
6	Rome, Italy	150,000
7	Antioch, Turkey	120,000
8	Syracuse, Italy	100,000
9	Ujjain, India	85,000
10	Datong, China	80,000
11	Tehran, Iran	80,000
12	Athens, Greece	75,000
13	Balkh, Afghanistan	75,000
14	Corinth, Greece	70,000
15	Anuradhapura, Sri Lanka	65,000

Table 2

	1500 CE	*Estimated population*
1	Beijing, China	670,000
2	Vijayanagar, India	500,000
3	Cairo, Egypt	400,000
4	Hangzhou, China	250,000
5	Tabriz, Iran	240,000
6	Istanbul, Turkey	200,000
7	Guar, India	200,000
8	Mexico City, Mexico	200,000
9	Paris, France	185,000
10	Guangzhou, China	150,000
11	Nanjing, China	150,000
12	Cuttack, India	140,000
13	Fez, Morocco	130,000
14	Edirne, Turkey	130,000
15	Xi'an, China	130,000

Table 3

	1900 CE	*Estimated population*
1	London, England	6,500,000
2	New York, U.S.	4,200,000
3	Paris, France	3,300,000
4	Berlin, Germany	2,700,000
5	Chicago, U.S.	1,700,000
6	Vienna, Austria	1,700,000
7	Tokyo, Japan	1,500,000
8	St. Petersburg, Russia	1,400,000
9	Manchester, England	1,400,000
10	Philadelphia, U.S.	1,400,000
11	Birmingham, England	1,200,000
12	Moscow, Russia	1,100,000
13	Beijing, China	1,100,000
14	Kolkata, India	1,100,000
15	Boston, U.S.	1,100,000

Table 4

	2020 CE	*Estimated population*
1	Tokyo, Japan	37,100,000
2	Delhi, India	26,300,000
3	Mumbai, India	23,700,000
4	São Paulo, Brazil	21,600,000
5	Mexico City, Mexico	20,500,000
6	New York, U.S.	20,400,000
7	Shanghai, China	19,100,000
8	Dhaka, Bangladesh	18,700,000
9	Kolkata, India	18,400,000
10	Karachi, Pakistan	16,700,000
11	Beijing, China	14,300,000
12	Lagos, Nigeria	14,200,000
13	Manila, Philippines	13,700,000
14	Buenos Aires, Argentina	13,600,000
15	Los Angeles, U.S.	13,500,000

Source: Adapted from Tertius Chandler (1987) *Four Thousand Years of Urban Growth: An Historical Census*, Lewiston: St. David's University Press; and United Nations (undated) *World Urbanization Prospects: 2005, 2010 Revisions*, http://www.un.org/esa/population/publications

FURTHER READING

Here, and throughout the book, I can only point in a few directions given the plethora of books published on all aspects of city life. A brief, broad, and historical overview of the latter, however, can be found in Witold Rybczynski's *City Life* (New York, NY: Scribner, 1995) and in much more detail in Peter Hall's *Cities in Civilization: Culture, Innovation, and Urban Order* (London: Weidenfeld and Nicolson, 1998). More culturally oriented are Raymond Williams's classic *The Country and the City* (London: Hogarth Press, 1985) and Max Weber's *The City* (New York, NY: Free Press, 1958). David Harvey explores more contemporary trends in market-oriented city development in his *The Urban Experience* (Baltimore, MD: The Johns Hopkins University Press, 1989) and Mike Davis gives an account of the plight of many an urban dweller today in his *Planet of Slums* (New York, NY: Verso, 2006). Finally, several recent edited volumes cover a wide range of cities and city processes on a global scale: Gary Bridge and Sophie Watson (eds) *A Companion to the City* (London: Blackwell, 2003); Richard T. Legates (ed) *The City Reader* (Hoboken, NJ: Taylor and Francis, 2011); Ronan Paddison (ed) *Handbook of Urban Studies* (London: Sage, 2001); and Stanley D. Brunn, Maureen Hays-Mitchell, and Donald J. Zeigler (eds) *Cities of the World: World Regional Urban Development* (New York, NY: Rowman and Littlefield, 2012).

There are many websites that are devoted to cities and city issues. Because these change frequently, the very select bibliography of each chapter will not attempt to name them or identify specific links. However, in general, the international organizations that maintain the best sites are the United Nations, the World Bank, the European Union, and the Organization for Economic Co-operation and Development (OECD). Other sites that look like they will last include Cities.com, ESRI Community Showcase, and GaWC (Globalization and World Cities).

NOTE

1 For my particular understanding and use of the term West, or Western, please consult the Glossary of the book. Also, several chapters have a case study box which treats the urban experience of other areas of the world, if rather briefly. More supplemental material of this nature, as well as on other city topics of interest, can be found at the publisher's website: http://www.routledge.com/books/details/9780415670791/.

CITIES AS THE SOURCE OF CIVILIZATION

The very existence of cities signifies the creation of a spatial division of labor between the city and the rural as a result of a social division of labor between those who work the agricultural fields and those who are able to do something else with their time. The key to the growth of cities and numbers of city people, then, very much depends upon both the ability of different societies to produce a surplus over and above day-to-day subsistence needs, as well as that of certain classes of people to lay claim to the surplus generated by other classes in the process of distribution. The specific nature of the earliest Ancient/Classical cities (to ca. 1250), in terms of social and built **morphology**, or spatially distributed characteristics, thus reflects the specificities of surplus production as well as the characteristics of which class came to control the distribution process.

In general terms, the social and built morphology of Ancient/Classical cities reflected and symbolized religious and political power, reinforced by military might. The most powerful city people in the earliest cities such as Ur and Uruk (ca. 3rd Millennium BCE) in the Tigris and Euphrates river basins of what is today the country of Iraq were religious or political figures. The same is true of cities of the Nile River Basin and the Yucatan Peninsula and other regions of early city emergence. The built morphology of these cities reflected this with religious and political monuments in the most central or otherwise most noticeable areas of the city in proximity to the habitations of the most powerful members of city society. Surrounding these monuments and habitations were the abodes of lower-class city people who catered to the needs of the city elite and who frequently provided the military force to sustain the existing social and spatial division of labor.

These characteristics of the earliest cities remained largely the same in the non-Western world until the emergence of Western intrusion after 1492. In the West itself, cities more or less exhibited such social and material characteristics until the end of the Middle Ages (ca. 1250). Indeed, at this level of generality, the most important aspect of cities and the city people who ran them throughout this very long historical time period was the ability to maintain control over, and even expand, the production and distribution of agricultural surplus. In terms of control, elite city people had to first coerce surplus production of the many non-elite rural producers who otherwise would be content to produce only for their own subsistence. This was done either via notions of religious obligation or by more secular, political forms of taxation for the most likely reasons of security and the maintenance of social stability. In either case, rural direct producers either were persuaded or, more likely, outright coerced to give up their surplus in exchange for religious and political services offered, or forced upon them, by city people. It is for this reason that some have characterized cities and city people as ultimately parasitic on their rural **hinterlands**.

CITY PEOPLE AND CIVILIZATION

Understanding these general characteristics renders it possible to determine why some human societies attained what most would call higher levels of civilization than others. The key is the generation and control of greater amounts of surplus and, then, the ability to support greater numbers of city people and the cities that house them. In terms of the first, it is possible to demand more and more surplus from traditional producers; but this is only possible up to a point. As this point is reached, producer resistance is likely to increase, which may overturn the whole arrangement in the form of outright social revolt. The more successful strategy, then, was to increase the spatial extent of the city's surplus generating hinterland by conquering more land and people for this purpose. So Ancient/Classical cities grew in population size and actual number to the extent that more and more hinterland was brought under the control of any given civilization. The Chinese, Persian, and Roman Empires, for example, included many cities and city people as a result of their respective imperial expansions over more and more conquered lands and peoples. All roads led to Rome, as they say, for two main reasons: to get the surplus from the ever-growing conquered hinterlands to the capital city and to allow the military might of Rome more efficient access for the forced maintenance, and expansion, of this very imperial hinterland.

Such increased surplus supported a very large population in Rome itself, as well as in many secondary and tertiary imperial cities in the empire as a whole. And this brings us back to this very notion of civilization. The Chinese, Persian, and Roman civilizations are considered to have been "great" by most historians because they were rich with innovations in language, numeracy, science, architecture, art, history, philosophy, etc. The key is that such innovations derive from city people who do not need to toil in the agricultural fields all day for their sustenance. City people can do other things all day and those other things lead precisely to what we call social innovations and the deepening of civilization. In short, the depth or profundity of early Egyptian or Roman (or Chinese, or Mayan, or Persian) civilizations were the result of the conquest over an ever-more expansive surplus-generating hinterland that allowed more and more Egyptian and Roman people to live in cities and do such specialized city things.

OTHER CITY PEOPLE

While it is true that all city people lived off the surplus generated by rural direct producers, not all city people enjoyed the seeming comfort and leisure that such an arrangement suggests. Of the total population of any Ancient/Classical city, the vast majority consisted of artisans/craftsmen, laborers, merchants, and low-level military personnel, both free and enslaved, who catered to the needs of the much smaller number of city elite. Taking note of this other city population is significant for two main reasons. The first concerns the activity and status of artisan/craftsmen and, especially, merchants in these cities. From the perspective of the twenty-first century, for example, it is difficult to conceive of such business people as having low social status within the city. Yet, in Ancient/Classical cities, this was the case as status was based mostly on heredity and the ownership of land, considered to be the source of all status and wealth in such agrarian societies. Indeed, the common behavior for such business people even well into the Mercantile/Colonial period (ca. 1250 to 1750) was to acquire landed property and therefore higher social status once they had attained a certain level of success in their trade.

Within this context, merchants were held by most people in particularly low esteem and even outright despised even by non-elites. Although merchants made available whatever exotic and other trade goods that the elites desired, the very act of trading was considered socially base or dirty, best left to those of equally base character, slaves, or foreigners. The reason for this is complex and varies with time and place, but can be characterized as arising from the consideration of how merchants behave and what their ultimate objective is. In terms of behavior, merchants must act on their own behalf in a relatively secretive manner. To be successful, they must keep their sources of supply, their route of trade, the amount they originally paid for the goods, etc., to themselves or only a very few close associates. These "trade secrets" are the very basis of their potential profits. For this reason, merchants appear (and many times actually are) untrustworthy to deal with, like the contemporary stereotype of a used automobile salesperson. One never really knows if the deal a merchant offers is good or even fair for that matter. After all, the merchant is only in it for the merchant's own benefit.

This last assertion points to the other main reason for the low esteem afforded merchants in most of human history up until the Western Renaissance (ca. 1250). Merchants do not trade for the benefit of society but for their own profit and social status. The good of the rest of society does not figure into this equation. In many civilizations, including the West until its Renaissance, such egotistical behavior was heavily frowned upon and, indeed, considered quite detrimental to the maintenance and further development of the civilization itself. In this respect, merchants were considered to be, quite literally, anti-social, a cancer on the health of the larger community.

For these main reasons, merchants were heavily regulated by traditional land-based elite most everywhere and certainly within Ancient/Classical cities. In fact, these two broad cultural reasons for anti-merchant sentiment were reinforced by another adverse sentiment more specific to this elite – that is, merchants, by their activity, actually pose a threat to the social power of traditional land-based elites. Their very ability to accrue personal profit on the basis of their own activity means that merchants can buy their way to social power and then vie for power with the traditional elite. From this perspective, merchants had to be regulated, and regulated heavily, as a specific means by which the traditional authority of the land-based elite could be maintained.

SOCIAL STABILITY IN THE CITY

It is important to dwell on this rendering of merchant behavior and ultimate regulation because it differs so much from today's common understanding. Indeed, why this difference seems so stunning to us now will be important for the story of the transition from Ancient/ Classical cities to Mercantile/Colonial cities in the next sections of this chapter. At this point, the other reason why it is important to recognize the other non-elite city people should be addressed. Quite simply, this majority, non-elite, city population of artisans/ craftsmen, laborers, merchants, and low-level military personnel ultimately had to be kept satisfied with their lot in city life in order for cities and, indeed, civilizations to maintain social stability. Now, "satisfaction" can be obtained by force exerted by the city elite, to be sure; but this is difficult to sustain as this minority elite would always have to watch its back, so to speak. Importantly, then, such satisfaction usually was obtained by other means – for example, what has been called with regard to cities of the Roman Empire "bread and circuses" or heavily subsidized cheap food and gladiatorial entertainment for the masses of non-elite city people. Such provision of bread and circuses gave such city people a sense of entitlement as legitimate members of the city population and, yet, in turn, diverted their attention away from their ultimate powerlessness within the city.

But the focus here is really on the very necessity of such an effort by, in this case, the Roman elite. Why did they even bother with such effort on behalf of non-elite city people? And here is where the emphasis must be placed on the nature of the city itself. The non-elite were not only the vast majority of the population of any given city, but they also lived and worked and recreated in very close spatial proximity to the elite and to each other. This spatial proximity allowed for the very rapid dissemination of information among the non-elite as they became socialized as part of a larger, densely packed, community of people of similar livelihood and life chances. Both of these truisms about city life, in general, rendered this non-elite population potentially a very real threat to traditional city elites. Indeed, this potential only needs an appropriate spark to make it a reality, such as the decreasing ability of the Roman elite in the last centuries of the Empire to provide enough bread and circuses as the costs of defending the larger empire became greater and greater. In short, the very nature of city life with dense populations on small territories renders cities potentially dangerous places for social elites, no matter how seemingly powerful they may be.

CHANGING CITY LIFE IN THE WEST

In most civilizations, such imminent dangers of city life were successfully regulated by the ruling political and ideological elite via both spatial segregation within the city and heavy regulation of non-elite, particularly merchant, activity as already noted. The social and spatial characteristics of Ancient/Classical city life thus continued for many centuries in Asia, the Middle East, Africa, and the Americas long after what has been called the "fall" of the Western Roman Empire during the period of 300 to 500. This fall of the "West," however, is crucially important for the present story. By 300, the Roman Empire had already been split into a western part, ruled over by the city of Rome, and an eastern part, ruled over by the city of Byzantium (today's Istanbul). Hence, the term "Western civilization" originally had a geographical connotation in addition to its current cultural one. While the Eastern Empire continued to enjoy relative social, political, and economic stability and city life remained largely in the Ancient/Classical mode, the Western Empire was besieged with invasions by so-called barbarian peoples of East-Central and Northern Europe. At the same time, the non-elite city people of Rome and other imperial cities were increasingly dissatisfied with their lot in life. As the invasions became more successful and as non-elite Romans became more demanding, cities in the West became more and more dangerous for traditional elites. As a result, many elites withdrew from the cities to their vast slave-labored country estates, the very source of their wealth and power in Roman society. And as cities continued to be attacked and overrun by non-Romans and their sympathizers, many other city people fled to the countryside as well.

Thus began the general ruralization of the Western Empire as Roman civilization increasingly merged with that of the ethnic groups of Central and Northern Europe, who traditionally led more rural, localized agrarian lives. Of importance is that the number of cities and city people declined dramatically during this period and for many more centuries to come in what is known as the Western Middle Ages (ca. 500 to 1250). This evolving European society thus lost the very people who create depth to civilization. Life for most Europeans revolved around a relatively few small landholdings grouped together near some means of protection from external attack, such as a large landholder's fortified estate or castle. Of the major institutions of the former Roman Empire, only the Catholic Church survived somewhat intact; but even the Church was now more rural, localized in fortified monasteries scattered across rural Europe. What city life did exist was on a much reduced scale both in number and population and overwhelmingly dominated by Church-based activity and authority. In this respect, the few cities that existed in the Western Middle Ages continued to exhibit the same characteristics as other Ancient/Classical cities, if on a much smaller scale, in general.

What was different was the geopolitical context of such cities. Hitherto, the most important Ancient/Classical cities were all regulated by powerful, centralized imperial authority over the civilization as a whole. With the disintegration of such centralized authority in the merging of Western Roman society with that of European peoples, whatever "cities" existed, existed in the relative isolation of whatever local authority was to be found, quite independent of other local authorities. In short, the political geography of Europe in the Middle Ages was one of relatively autonomously organized, much smaller political territories surrounded by no little disputed or even entirely unclaimed land.

Such gaps in identifiable and enforceable political authority were important for two broad reasons. First, they signified territorial resources and sometimes peoples that were potential conquests for those authorities which did exist. In other words, the localized political authorities of Middle Ages Europe were potentially, and often actively, engaged in competition to acquire such areas and people in order to become more powerful. The greatest minds of the Middle Ages therefore came to be focused largely on military pursuits, both offensive and defensive, as well as organizational and technological. This martial competition led to much innovation, in fact, which became important both for the eventual development and protection of European cities, and also for the European exploration and conquest of most of the rest of the world's resources and peoples after 1492.

Second, highly decentralized political authority rendered it extremely likely that individuals or whole groups could successfully resist, or even escape from, political authority altogether. Indeed, this possibility will be important to our specific story of the later city-led Renaissance period of Western history (ca. 1250 to 1650). In the end, the decentralized geopolitical structure of Middle Ages Europe represented an essential context by which political power became, by necessity, a negotiation between the more and less powerful, whether at the geographical scale of Europe as a whole or within these regional polities themselves, between emerging cities and growing numbers of city people and traditional land-based political authorities.

THE RE-EMERGENCE OF CITIES AND CITY PEOPLE

Nevertheless, the real significance of this geopolitical structure of the European Middle Ages for cities and city people only truly manifested itself toward the end of these so-called Dark Ages. Throughout most of this period, the relatively few, sparsely populated, towns and cities that existed in Europe largely retained the characteristics of other Ancient/Classical cities in that their main function was administrative and ritualistic. The built and social morphology of these cities reflected this, with Catholic cathedrals and the higher clergy located in the central areas largely symbolizing and actually functioning as both administrative and ritual authority. Reflecting the highly regional and local nature of European geopolitics, there were thus relatively few full-time city people other than those who catered to the needs of the Church and its administrative activities. This meant that most major cities of the European Middle Ages remained quite small and singular in function. The vast majority of people lived in even smaller villages or towns, mostly centered around a large landowner's mansion or castle. By 1000, for example, Rome, then the largest city in Catholic Europe, had a population of only around 25,000 living among ruins left by a population once as large as 150,000 or so in the year 150. At the same time, the population of Constantinople (formerly Byzantium) in the East was about 750,000, and even Cordoba in Arabic Spain had a population of as much as 200,000.

This is the real reason that this was a "dark" age. It was a time in which the West lost its cities and city people, the very people who deepen civilization, as already described. The great majority of people in this time period were busy toiling in the agricultural fields all day for both their main sustenance and to pay the exorbitant amounts of rent due to the owners of the land they worked. The agricultural surplus of localized workers was not enough to support much more than this because there was little incentive for the semi-slave "serf" workers to produce more efficiently, as any more surplus would simply be siphoned off as rent by the local landowner. There was also the more general uncertainty of production in a time of regular conflict and invasion. Again, there was no larger, more centralized geopolitical authority which could siphon off surplus from a greater territory while, at the same time, enforcing relative peace within such an empire.

The Dark Ages were dark, then, because of the relative loss in the West of its cities and city people. But such Darkness was never complete and by the end of this period (ca. 1000 to 1250) there were three emergences significant for cities that were of growing importance. The first was the spread and growth of city-based, Church-sponsored "universities" in which increasing numbers of scholars and their students debated Catholic doctrine and, increasingly, the relationship between Christian dogma and the ancient pagan literature, particularly that of Plato, Aristotle, Virgil, Cicero, and Ovid slowly becoming known again in the West. Such a concentration of scholars and learners in cities provided the basis for an even larger mass of city-based people who would eventually come to question the traditional beliefs of society as a whole. The second emergence was the growth and expansion of city-based and other periodic regional market fairs in which merchants and producers from across a larger territory would come together to sell their goods and agricultural surpluses. Such market fairs, as they continued to grow, added an additional, increasingly major, function to the cities of the Western Middle Ages. Cities became, in some cases, central market sites as a result of the periodic market fair itself becoming more and more permanent in specific places such as Frankfurt and Lyon.

The third emerging trend of this time period was the steady increase in the population of Europe. For reasons not quite fully understood, it is estimated that total population actually doubled in Europe between 1000 and 1300. This relatively rapid increase in population is significant because it put serious pressure on agricultural producers to produce more surplus to feed the growing number of people in their midst. This led both to the taking into cultivation of more land, as well as to innovations such as the three-field system of cultivation, the use of horses in cultivation, and the four-wheeled cart. Of most interest is that this growth of population made it increasingly costly to enforce traditional regulations on the productive activity and the movement of people. It also provided a growing market for surplus agricultural production, drawing increasing numbers of producers into the marketplace. Finally, to the extent that opportunities for this growing population to eke out a livelihood in the countryside remained relatively stagnant in most cases, much of this so-called "surplus" population, particularly of the lowest, most vulnerable classes, migrated to cities as the only possibility left for actual survival.

In the later Middle Ages, then, cities and towns began to grow and diversify in function and population, including an emerging merchant class as well as a growing lower-class contingent. Aside from the general increase in population, the growth of local commerce and longer-distant trade was the most important cause of such city growth. This change in the function and population of cities in the West was hastened considerably by the great Western Crusades of the eleventh and twelfth centuries to take back the Holy Land in Palestine for Christendom. These Crusades eventually resulted in the increasing ability of Europeans to move people and goods over larger distances. They also brought more and more European people into contact and ultimate relation with each other. The Crusaders also greatly increased demand for the services and goods of the merchants, shipbuilders, and fleet owners of the port cities of Venice and Genoa in northern Italy, from which they departed for the Holy Land. Merchants in these cities profited tremendously as a result, allowing more and more economic opportunity for growing city populations. In turn, those Crusaders who actually made it to the Holy Land encountered (and also plundered) many exotic goods, such as fine silk and cotton textiles, elaborate ivories, and exotic spices which, when brought back to Europe, engendered increased demand and profit in the marketplace, particularly among the elite. Finally, the extremely expensive Crusades rendered the many traditional landowners who took part, as well as the Catholic Church itself, heavily in need of financing initially and then ultimately laden with debt afterward. This provided much leverage over these traditional authorities on the part of those city-based bankers and money lenders who were able to extend such financing.

THE INDEPENDENT MERCHANT CITIES OF THE WESTERN RENAISSANCE

As trade and overall commerce continued to grow in Europe, the cities of the late Middle Ages began to take on quite novel characteristics. Particularly in northern Italy, in Venice, Florence, and Genoa, and in Flanders (today's north-western France and western Belgium), in Antwerp, Bruges, and Ghent, the main function of cities was becoming trade and city life was increasingly dominated by merchants, bankers, craft producers, and a growing urban working class. As such cities prospered, more and more opportunity for economic advancement became available to others migrating to them. Even many traditional land-based elites in proximity to such emerging cities began to engage in commerce and trade as it became increasingly clear that such activity potentially increased wealth and then power. Those who did not participate themselves also got caught up in the new city-based mode of production either through the accumulation of debt, as mentioned above, or by extending financing to those more actively involved. Trade and commerce thus became a new means for many of the traditional elite to increase their competitive position with regard to their rivals in other polities on the basis of their increasing wealth and, then, power.

Precisely because of these latter possibilities, many land-based authorities began to cultivate their relationship with city-based merchants and bankers by extending them more and more freedom from traditional political regulation. The growing belief was that the cash cow of the city was best left alone to find its own fertile pasture to fatten itself. At the same time, those land-based elite most indebted to city-based people were most willing to curry favor with the latter for better financial terms or even forgiveness of debt in return for more freedom from traditional authority. Finally, and perhaps most importantly, as city-based people became increasingly more prosperous on the basis of growing trade and commerce, these people themselves were increasingly able to buy their own freedom from traditional political authority, either peaceably or, indeed, by force, by financing their own soldier-mercenaries and military technology and organization.

NEW CITY-BASED MERCHANT CULTURE

At this point it is necessary to pause and reflect on the evolution of this new type of city to make better sense of its very novelty. As European merchants became richer and increasingly secured themselves behind city walls with self-financed mercenaries, they became more and more independent of traditional authority altogether. To the extent that they were successful at this, cities such as Venice, Florence, Milan, Antwerp, Ghent, and Bruges became increasingly powerful European polities themselves as city-*states*. And, yet, very much unlike the Ancient/Classical city-states of Asia, America, Greece, and Rome, these new city-states were organized by a class of people who arrived at their power not by heredity or ordination but, rather, by their own activity and success in the commercial marketplace. They were, in this sense, self-made in their social power and their cities were self-organized for mutual protection. The merchants of the European Renaissance thus increasingly broke free from the traditional regulations that severely limited their numbers and their activity in other regions around the world. These were new city rulers who would go on to rule their cities in an entirely new way based more on the exigencies of merchant culture than on tradition or ritual.

The Renaissance – or "rebirth" – of the West was thus a recovery of cities and city life, to be sure, but these were very different sorts of cities and city people than existed in the Ancient/Classical period and in the rest of the contemporary world. To understand the nature and ultimate significance of this difference, it pays to explore the main characteristics of this evolving merchant city culture. First, as alluded to above, to be a successful merchant it is necessary to act as an individual in a private way for thoroughly self-serving reasons. Second, the focus of this activity must remain continually on material profit. Third, because of the first two exigencies, merchants must always be vigilant as to their competitors and this "keeping up with the competition" is an important impetus to innovation in the products and process of trade. Fourth, this latter exigency also involves seeking efficiency with regard to time, as time is money, so to speak, in getting goods to market and sold. Fifth, as the overriding goal is material profit in the marketplace, most anything can be considered as a possible saleable commodity, even services such as reading, drawing contracts, accounting, dramatic performances and other artistic works. Sixth, literacy and numeracy are generally necessary to be a successful merchant in order to keep better track of business transactions, contracts, and inventories.

Of course, this list of the main characteristics of merchant culture is hardly definitive and surely too general. But it does underscore that this new, emerging city-based merchant culture was quite different from the tradition-based agrarian culture from which it sprang. Merchant culture is all about initiating innovative change and pursuing material accumulation, growth, and expansion on the basis of private, competitive individual activity. Importantly, because social power and status could be gotten by success in the growing commercial markets of cities, there was an extremely strong incentive for more and more people to adopt these cultural characteristics. It also, for this very reason, provoked a growing fear of cities and city people on the part of most traditional land-based authorities, both secular and ecclesiastic. According to this view, cities increasingly were the places where new socially dangerous classes were located and, even more menacingly, these classes were in control of their own affairs as rival powers in the competitive, decentralized geopolitics of Renaissance Europe.

More positively, city-based merchant culture brought forth much of what is usually considered the grandeur of the civilization of the Western Renaissance. It was, for example, largely the change in patronage of the arts from the Catholic Church to merchants that encouraged innovations in artists' techniques and presentation. Artists themselves entered into merchant culture as they competed with each other for such patronage, experimenting with both their process of production and the ultimate product. The very fact that artists began to sign their work to distinguish it from that of their competitor artists attests to the significant changes taking place in the cities. The same cultural characteristics infiltrated other sectors of Renaissance society, including architecture, literature, political advising, and scholarship, in general, as merchants became paying customers in emerging markets. Just like artists, expert purveyors of such products began to compete among themselves, leading to much innovation in these sectors as well. In short, the vaunted Renaissance rediscovery and embellishment of *humanism* (focused, as it was, on the possibilities of individuals) and *realism* (focused, as it was, more on reality as it really is) can both be directly related to the individual merchant's quest for a better life in this world rather than waiting for reward in the afterlife. And it must also be emphasized again that both of these persuasions of city-based merchant culture were considered by many to be dangerously contrary to tradition, even quite heretical.

THE REFORMATION AND CITY LIFE

For all the reasons outlined above – the fall of centralized imperial regulation and the decentralized nature of European geopolitics; regular warfare and the competitive pursuit of political power; population growth; the emerging self-gotten political freedoms of cities; the long-distance linkages of the Crusades; and, one can add here, regular plagues and pestilences which killed both lord and serf randomly – the new cities and city people of the Renaissance rapidly grew in numbers and spread throughout Europe. The pull of city life and the freedoms that city people enjoyed was a powerful magnet for those, particularly of the lower classes, who wished to better their own social status. It was also a powerful incentive for the self-protection of these freedoms from attempts by the land-based elite, whether secular or ecclesiastic, to reassert their traditional political authority over city people.

So the new merchant authorities of these cities created their own governments within fortress walls protected by soldiers of their own employ. But another sort of protection was necessary as well. This was a protection against the ideological admonishments of the still dominant Catholic Church against things such as profit-making on the backs of other good Christians as a means to better one's life on Earth, as well as the improper charging of interest on loans. Fortunately for good Christian merchants, another Renaissance event, the **Reformation** (ca. 1500), and the rise of Protestantism, loosened this intellectual hold on emerging city-based merchant culture. Significantly, the Protestant movement took off fastest in the cities of Northern and Western Europe, where the merchant class was growing most rapidly. More significantly are the reasons why this was the case. In arguing that the Catholic Church had lost its way through wealth and corruption at the highest levels, Protestants argued for a return to what they considered to be the real teachings of Jesus outlined in the Holy Scriptures. In short, the only thing necessary to be a good Christian, according to this new Protestant teaching, was an individual's faith and obedience to scripture.

From this time onward, then, it was possible to be a good Christian even in solitary meditation of the Holy Scriptures. This is important for the following reasons. First, it meant that it was no longer necessary to give elaborate monetary tithes and offerings to the Catholic Church (or any other church) in order to have its official clergy intercede on one's behalf before God. That meant one could be a good Christian and still accumulate one's own profits. Second, because knowledge of scripture was so important in this new Christian teaching, most Protestants were instructed to become literate in order to read and meditate over the scriptures themselves. This was entirely against Catholic teaching of the time which cautioned against this for fear of misinterpretation of these very scriptures. Before the Reformation it was, indeed, an actual sin to translate the Bible from the original Latin or Greek into the vernacular tongues. It was better to be illiterate, according to this view, than to be led astray by the misinterpretation of scripture.

Since city people were generally more literate than the rest of the population, either as scholars and academics or as merchants and bankers, it is no wonder, then, that the Protestant message was absorbed so quickly in cities. Indeed, it was the Reformation that secured the fame of Gutenberg's newly invented printing press throughout Europe, as demand increased rapidly for Bibles translated into German, French, Dutch, English, etc. But it was a specific teaching of Protestantism that fully loosened the hold of Catholic admonishment on merchant behavior. This was the notion of predestination: the idea that God, knowing all and everything, already knew whether one was going to either heaven or hell before one was even born no matter how one conducted one's life on Earth. Briefly, while this teaching seems quite dismally fatalistic, Protestant authorities argued that, while this may be true, if one does, in fact, lead a successful life including, significantly, becoming wealthy, it probably signals that one is, indeed, one of the "elect" bound for heaven.

So now one could be a good Christian by being a good merchant with no holds barred except to not be overly prideful and wasteful of one's wealth. This fully freed city-based Christian merchants to follow their individual path toward riches and power. In fact, it was a very strong incentive for all to follow this path actually to prove elect status. Because of this, the Catholic Church reacted by enforcing its orthodoxy more rigidly in those parts of Europe where it remained most powerful, such as southern France, Italy, Spain, and Portugal, which all went through what is called the Counter-Reformation. This included severe crackdowns on the dissemination of written documents and new city-based freedoms and merchant activities. Cities and city people all became suspect in these regions of Europe as land-based political authority reasserted itself fully backed by newly rigid ecclesiastic teaching.

CHANGING LOCATION, FORM, AND FUNCTION OF MERCHANT CITIES

These new, rapidly emerging merchant cities of the Renaissance were significantly different from earlier cities in many ways. Cities now came to be located and developed much more for economic reasons as opposed to religious, political, or military considerations. As transporting trade goods was always cheaper over water, for example, coastal ports and riverfront sites became favored locations. In addition to the Renaissance Italian and Flemish cities already mentioned, this period saw the rise of major cities such as Amsterdam, London, and Stockholm, and more inland linked river-based cities such as Lübeck, Hamburg, and Cologne. As Europeans began to explore the wider world, this coastal emphasis became even greater at the same time that the main focus of trading activity shifted from the eastern trade via the Mediterranean Sea to a western and southern orientation via the Atlantic Ocean.

The social and built morphology of Mercantile/Colonial cities was also different from earlier cities. As economic activity became the focal point of city life, merchants and merchant warehouses, bankers, money lenders and their establishments, and commercial artisans and their shops were housed and did their business in more central areas of the city. The most prominent built structures of such cities thus came to symbolize the new power and prestige of this new economic elite. The rest of the built environment was reconstructed or newly constructed to meet the needs of business. Each city thus increasingly came to be constructed around what today is called a **central business district (CBD)**, which, during this time period, also included the habitations of the most successful merchants and bankers.

So it was that changes were taking place at this time in the location, form, and function of Western cities on the basis of primarily economic considerations. At this point, there are two further issues of significance concerning these changes. First, tied as they were becoming to economic success for survival, the fate of cities was largely determined by the ability of city authorities to keep up with the latest innovations in transportation, products, and, eventually, the process of production. While this latter will become important only later, it is clear that as the **mode and network of transportation** comes to change, so does the economic fate of particular locations, as well as the internal morphologies of cities. For example, as modes of overland transportation changed from horses to horse-drawn carts to railroads to automobiles, and as new networks of highways, train tracks, and freeways emerged, the overall economic fate and characteristics of earlier water-based ports changed, sometimes dramatically. Changes in transportation also shifted the center of business, in many cases, away from the actual port area to other parts of the city or, indeed, to other cities altogether. Similarly, as dominant consumer tastes changed from, say, woolen or linen clothing to that of cotton, those cities which continued to concentrate on the production and trade of the former lost competitive edge, and therefore economic viability, to those who switched to the latter. In short, as cities and city life became increasingly based on economic activity and viability beginning in the Renaissance, the fate of each became ever after more a matter of competition among all to keep up with the latest innovations in trade and, eventually, production.

The second significant issue concerning the rise of economically based cities is social and cultural. As mentioned, the new city-based authorities of the Renaissance were merchants and bankers, self-made individuals on the basis of their business prowess. In these new cities, what mattered was how successful in the market one was, not just what social class one was born into or how large one's inheritance might be. This freed up everyone, ideally, to participate in the market; to think up better ways to do things, to eventually build better mousetraps and so on. This certainly will be the basis of much innovation emanating from cities to be sure; but, equally important, it signifies both active participation by individuals in their own destiny in life and a significant enhancement in the cultural status of merchants. The highest social status could now be obtained by anyone via successful profit-making in the commercial marketplace. This was a formidable incentive for all to participate and, as such, it came to provide an engine for continual market expansion as more and more people became engaged in commerce. Finally, because individuals were now able to make their own social status based on their own activity, it is no wonder that they also came to believe that they should also make their own rules by mutual agreement as to how they were to be governed or otherwise conduct their lives. This rudimentary form of what we call *democracy* in Renaissance cities was what was considered most dangerous by many traditional land-based authorities of the time. If such ideas and reality were to spread, it posed a direct threat to the very existence of their power. This simply reinforced the notion, common among this traditional elite, that, by their very nature, cities were the breeding ground of social trouble.

THE GLOBAL EXPANSION OF CITY-BASED MERCHANT CULTURE

The implications of these last points will be discussed in more detail in later chapters. Here it is necessary to point out that this city-based merchant culture expanded quite rapidly both within Europe and around the world via Western imperialism. In Europe, the continual growth of commerce and trade increased the size of existing cities and brought more cities to commercial life. A major reason for the latter was the movement of merchants from region to region in search of further opportunities for profit. Merchants migrated to, and settled in, cities across Europe, regardless of their original language and culture. Italian merchants from Genoa and Venice were particularly active in this way, settling and doing business all over Europe and even farther east in port cities along the Black Sea and in North Africa. Of importance is that this spread of city-based merchant culture provided significant impetus to the further expansion of Europeans across the globe.

It was to out-compete the Italian city-states within Europe and to completely bypass the infidel Islamic merchants of North Africa and the Middle East, for example, which drove Portugal and then Spain to set off in search of sea routes to the very source of the luxury commodities and rich spices of Asia. As is well known, by the late 1400s, Portuguese sailors had already successfully rounded the tip of Africa and entered the Indian Ocean and, of course, the Spanish had found what was to them a "new world" across the great western ocean. The importance here of these European voyages and ultimate conquest of non-European peoples and territories is essentially threefold. First, the new sea-based intruders from Europe very rapidly disrupted traditional indigenous trading patterns and relationships by muscling in with their superior naval war technology, honed, as it was, in brutal competition with their fellow Europeans back home. Very soon, Western merchants were plying the trade among Asian ports as well as back and forth to Europe. Second, as more and more Europeans entered Asian waters, there was a significant increase in demand for particular tradable goods. The first intrusion dramatically decreased the profits of local indigenous traders, leading to internal economic stagnation, while the second brought both increasing specialization and commercialization of indigenous production now increasingly for export to Europe.

The third result of formal European intrusion into the non-European world is intimately tied to the first two and of utmost importance. The new European presence essentially reoriented traditional indigenous trade patterns from internal circuits to sea coasts. Sea-based European demand for slaves in Africa, cotton textiles and spices from India, silks and porcelain from China, and spices from South-East Asia resulted in the growing importance of coastal trade *entrepôt* cities such as Bissau, Goa, Macao, Malacca and so on. Similarly, in the fully conquered territories of the Americas, the European quest was first to demolish and reoccupy the former imperial cities of the Aztecs and the Incas and then to get American goods out and on to Europe, whether it was sugar and coffee through São Paulo, or beef through Buenos Aires, or gold through Vera Cruz, or silver through Cartagena, or cacao through Caracas, and so on. The point is that, from the very beginnings of European overseas expansion, city development in the non-European world began to take on an externally focused, largely coastal aspect that remains very much the case today.

Box 2 From Babylon to Teotihuacan

The first major cities in human history were found mostly in non-Western areas of the world, as clearly shown in the population data presented in Box 1. The earliest of such cities, many argue, arose in Mesopotamia, roughly an area covering the modern countries of Iraq and northern Syria. Intense agriculture emerged in the fertile environment of the Tigris and Euphrates river basins in the southern part of this region and the Sumerian cities of Ur, Eridu, and Uruk evolved as early as the mid-fourth century BCE. By the mid-third century, a system of city-states linked by trade and politics existed from the Persian Gulf to the city of Babylon, which was located near the modern city of Baghdad. Because of this, many argue that this area of the world was the most urbanized of ancient times and, thus, the very cradle of human civilization. Indeed, the wonders of the great city of Babylon are the stuff of historical lore, from its massive ziggurats and temples to its large water-laden gardens, which may or may not have been hanging throughout the city. It was so large relative to other cities of its time that it is said that Aristotle (384 to 322 BCE) did not think it was actually a city but, in fact, an entire nation. Finally, similar ancient urban systems came to exist in China from the Yellow River to the Yangzi Basin, leading directly to the marvelous deepening of Chinese civilization that came to attract the attention of early European explorers from Marco Polo in the thirteenth century CE to Christopher Columbus and his imitators from the fifteenth century CE on.

On the other side of the globe, great city populations also emerged in Mesoamerica, although much later in human history. In what is called the Basin of Mexico today (a fertile highlands plateau in central Mexico surrounded by volcanic mountains which is also called the Valley of Mexico), the great city of Teotihuacan was established in 100 BCE and continued to grow in size to the 300s CE. It is believed that this city, the population of which has been estimated to have reached as much as 100,000 at its height, survived as late as the 600s or 700s CE before being rather mysteriously destroyed and depopulated. The size of this great Mesoamerican city can still be discerned by its monumental ruins, including large-scale pyramids, grand avenues, and even commoner dwellings, that continue to exist at its site some 50 kilometers (30 miles) north-east of today's Mexico City. On a smaller scale, the Mayans located in today's Yucatan Peninsula and Central America developed city-like ceremonial sites from the third to the tenth century CE and the Incas of the South American Andes and Pacific Coast developed cities beginning in the thirteenth century CE, including the imperial administrative center of Cuzco and what some call the religious site of Machu Picchu in today's Peru.

FURTHER READING

A classic book concerning the history of city life is Erwin A. Gutkind's multivolume *International History of the City* (Glencoe, MN: Free Press, 1964–1972). Although somewhat dated, this anthropologist's work maintains a very broad yet focused approach to the global emergence and evolution of cities. A couple of books spend more time detailing the physical environments of Ancient/Classical cities: Spiro Kostof's *The City Shaped: Urban Problems and Meanings through History* (London: Thames and Hudson, 1991) and Gideon Sjoberg's *The Pre-Industrial City: Past and Present* (Glencoe, MN: Free Press, 1968). Aiden Southhall attempts a much more ambitious historical account of both the physical and social environments of cities throughout history, particularly in the Global South, in his *The City in Time and Space* (Cambridge, UK: Cambridge University Press, 1998). Finally, Marc Van de Mieroop provides details about Ancient Mesopotamian urbanism in his *The Ancient Mesopotamian City* (Oxford, UK: Clarendon Press, 1997).

3

FROM TRADING TO INDUSTRIAL CITIES

The emphasis in the last chapter on the emergence of merchant culture in the West runs throughout this book. As it spread throughout Europe and eventually to European imperial possessions abroad, this city-based culture basically freed up more and more people to achieve social status on the basis of their own activity. This was a powerful incentive to take part in commercial activities where this was possible. It was also the source of much innovation in the products, methods, and organization of trade as new ways of doing things emerged as increasing numbers of people competed with each other for success in the marketplace. In the mercantile period (ca. 1250 to 1750), the increasing commercialization of Western life entailed two major, and quite linked, evolutions. The first was a growing **social division of labor**. As more goods became available for sale in the marketplace, it became less and less necessary for families to supply all of the goods necessary for their own livelihood. By selling whatever surplus they had produced of one good – say, poultry or eggs – in the marketplace for money, they could buy other necessary goods – say, milk, or shoes, or candlesticks – with the money that they made. This allowed families to concentrate more on what they wanted to produce, or what they produced best, and then increasingly to rely on the marketplace for other things they may have needed.

Of importance is that this growing specialization of family production took place most rapidly in proximity to the growing numbers of merchant cities of the Renaissance. It was, in this sense, a city-generated phenomenon. The second thing to notice about this growing social division of specialized labor is that it generated no little innovation in the production process. As more and more people concentrated on the production of goods they were best at producing, they inevitably found even better ways of producing these goods. They certainly had the incentive to do this as the more surplus that they produced for sale, the more money they could use to buy other necessary goods for survival. Not only that, but the better and better they became at producing this surplus, the more supply of this particular good – poultry, eggs, shoes, textiles, etc. – became available in the commercial market, generally resulting in lower prices for each. Lower prices, in turn, brought more and more people into the marketplace as they did not have to sell as much of their own surplus to acquire the goods that they did not want to keep producing. In short, the very spread of commercialization, if it is allowed to continue, sets forth a self-propelling, inherently expansionary process that essentially feeds off of itself.

But the key phrase is "if it is allowed to continue." The historical paradox is that the early merchant activity of Spain and Portugal came to be suppressed by continuing heavy political regulation well throughout the Renaissance period. Even though these countries came to control the first major overseas empires of Europe which provided them with tons and tons of American silver and gold, and relatively vast amounts of spices, cotton and silk textiles, and porcelains from Asia, Spain and Portugal were not able to convert this early wealth and power into lasting dominance within Europe. The treasure and goods that flowed back to the port cities of Seville and Lisbon eventually flowed right through Spain and Portugal to the growing numbers of self-governed merchants in northern cities such as Antwerp, Amsterdam, and, increasingly, London.

There are many reasons for this, of course, from the fact that the overseas explorations and conquests of Spain and Portugal were undertaken not by merchants themselves but by agents of the Crown for the Crown to imperial overstretch and armed conflict within Europe. Yet surely the continually heavy political regulation of city-based merchants in Spain and Portugal itself played a key role. Add to this that both countries implemented quite rigid Counter-Reformation policies, including the famous Inquisitions, in their attempt to purify the Catholic faith and it is relatively clear that neither Spanish nor Portuguese culture was transitioning toward the full city-based merchant culture of the Renaissance at the time. In fact, a good many merchants operating within the two countries actually migrated or were exiled northward to the growing cities of the north for this very reason.

TOWARD AN INDUSTRIAL REVOLUTION

Thus, cities such as Antwerp, Amsterdam, and London became increasingly wealthy and powerful in addition to the Renaissance forerunners in Flanders and Italy. In fact, global trade was shifting in importance toward the Atlantic coast, as European ships were now plying all of the world's oceans. It was no longer just the Spanish and Portuguese, but all the emerging powers of Europe, and particularly the Dutch, French, and English, who were attempting to get a piece of the action of overseas exploration and trade. The competition for power within Europe both accelerated these overseas journeys and greatly increased the amount of overseas goods available in European markets. Of particular importance for the growth of internal European trade was the vast increase in the amount of gold and particularly silver that was coming in from the Spanish and Portuguese empires in America. This rapid increase in the money supply greatly facilitated the growth of internal commerce by providing an increased means by which to convert the surplus production of any good into money profit in the marketplace. This, in turn, encouraged more and more European producers to engage in trade, thereby both increasing the supply of saleable goods and leading, again, to increasing specialization of production within an enhanced social division of labor.

The key at this point is this notion of "production." As increasing numbers of people were entering the market to buy these everyday goods that they no longer were producing themselves, others were thinking about how best to meet this demand. Of course, city people by the very nature of their non-agricultural existence were particularly needy in this commercialization of the spatial division of labor between cities and their hinterlands (now global in scope).

But the point is that this trade in goods was increasingly generalized to everyday necessities that were actually being produced by more and more local people. In other words, it was no longer just a trade in luxury items wholly produced elsewhere for the few wealthy who could afford them. As a result, many more minds were now set on determining how to substitute, for example, locally produced textiles or locally produced ceramics for those from overseas. How best, in other words, to use local raw materials and labor to produce similar goods to those now traded in the luxury market.

In this respect, industrialism is essentially an extension of city-based merchant culture to the realm of actual production. If a profit could be made by actually producing whatever good is in demand, then there were those who would set off in this competitive pursuit. And, indeed, this pursuit led to many innovations in products such as everyday ceramic dishes for what is still called "fine china," everyday cotton wear for elaborate calico prints, etc. It also led to innovation in production processes such as factory production for artisanal handicraft as well as in transportation, both in terms of mode, with railroad cars replacing horse-drawn wagons, and network, with rail lines replacing horse and buggy paths. Specifically, what has become known as the European **Industrial Revolution** (beginning ca. 1750) involved major innovations in three areas important for the present account of city development: the supply of productive power, the organization and technology of production, and the modes and networks of transportation.

POWERING NEW FORMS OF CITY-BASED PRODUCTION

The first innovation of note for cities of the industrial age involves the *supply of productive power*. During this period, there was a transition from animal and water power to coal-based steam power and then on to electricity and petroleum. Increasingly, steam power, in addition to providing readily available, powerful force, freed the locational requirements for handicraft and eventually factory production from waterways. It also provided the power for the new transportation mode of steam locomotives that revolutionized overland transportation as they ran on new and continually expanding networks of fixed rails. This greatly expanded the resource and market hinterlands of cities. It also dramatically changed the economic efficiencies of city location and the built and social morphology of industrial city life. New cities such as Manchester, Charleroi, and Lille emerged as inland mining and industrial sites. In turn, within major cities, railroad stations began to focus increasing numbers of people and businesses on particular central sites, and railroad tracks represented major dividers of neighborhoods. Similar changes will eventually take place as a result of further innovations in the productive power supply based on electricity and petroleum, as we will see later in this chapter.

The second major innovation of the Industrial Revolution important for cities involves the *organization and technology of production*. Part of this has already been dealt with in terms of the maturing social division of labor of increasingly specialized production. Here, the emphasis is on the way in which goods were produced. Before industrialization, goods for trade were made either by very skilled artisans such as cobblers, button-makers, furniture-makers, etc., and their apprentices in small shops or by less skilled producers, yarn spinners, and weavers in their separate rural cottages. The advantages of artisanal production are that the direct producers can limit the amount of goods that are sold – by limiting the number of skilled producers via restrictive apprenticeship rules – thereby maintaining high prices for their products. Notice, however, that this is also the disadvantage of artisanal production: a limited amount of goods at relatively high prices. Similarly, the advantage of cottage production (sometimes called the "putting out" system of production as in merchants putting out raw materials to the cottages for final assembly) is that it is relatively cheap in labor cost for the amount of material, usually woolen, linen, or cotton textiles, produced. The disadvantage of such a production process is the unevenness of quality and timing of the output from each cottage and the inability to enforce more uniformity and control in this regard.

This organization of production changed dramatically as part of the industrialization process. First, highly skilled artisanal production gave way to a process that split the production of finished goods into discrete productive steps, each one of which was performed by a different worker. This form of production was described and championed very early in the industrial period by Adam Smith (1723 to 1790), considered by many to be the first modern economic thinker. In his famous book *The Wealth of Nations* (1776), Smith describes this process with respect to the production of stick pins, and it is worth dwelling upon this description in order to understand the full implications of the division of labor in the production process that remain the case to this very day. In the mercantile period, for example, stick pins (used to fasten clothing and ornament hats) were made by artisans from the raw metal to the finished product. Smith argued, however, that a more efficient way to make pins would be to engage many laborers in a sequential process of production. One laborer could, for example, draw out the wire from the metal, another laborer could straighten the wire, another cut it, another sharpen it, another ground the top for placement of the head, etc., in discrete steps toward finally packaging and selling the pin by still other laborers. A modern example of such an **assembly line** of production would be automobile manufacturing, with one person doing the chassis, another doing the wheels and tires, another the windshields, etc., all day long. Adam Smith's point was that this type of production process was much more efficient in the sense that, at the end of the day, many more stick pins would be produced per worker than could possibly be produced via the old artisanal way. This meant that the prices of such pins would necessarily fall given competitive market processes, and then more and more people would be able to buy them instead of make them themselves or do without.

This new assembly line process of production eventually became more and more established in all sectors of industrializing countries. And this had enormous implications for city life, particularly in terms of its effects on workers. Effectively, the skill of making stick pins (and much else) was taken out of the artisan and put into the production process itself. In other words, each individual worker was now only responsible for, and then only came to really master, one stage of the production process, such as drawing the wire from bulk metal. This particular worker no longer had to know how to produce the entire pin as a finished product. The only skill that this worker needed was that of drawing wire, a "skill" which could be acquired in very little time on the job itself. Importantly, the value of each individual worker for the production process as a whole thereby dropped considerably. Workers became, in this sense, easily and, more importantly, much more cheaply replaceable for this very reason.

DESKILLING THE NEW CITY-BASED FACTORY WORKER

This new manner of industrial production thus entailed a rapid increase in the demand for unskilled laborers who were just then migrating in good number to cities as a result of population growth and a lack of opportunity in rural areas. These already poor, propertyless people began to fill up extensive, extremely densely packed, increasingly **slum**-like neighborhoods in industrializing cities. Before moving on, however, it is important to note one further result of this division of labor that will be important over time. Given the relative simplicity of the work at each individual stage of the production process, not only was it easier to find a cheap, unskilled worker to do the work, but it also became that much more a possibility to dispense with the need for human labor itself. A simple machine might be able to do the simple task just as well and machines never call in sick or go on strike. And, sure enough, in the very early years of the Industrial Revolution, mechanical innovations in weaving textiles (the flying shuttle) and spinning thread (the spinning jenny, water frame, and mule) appeared, putting many a hand weaver and thread spinner out of work altogether.

The details of these sorts of industrial innovations of the time can be found in any general history textbook. Of note here is that such innovations are made possible and, indeed, probable for the following three reasons important for the current story line. First, in addition to what has already been said, the division of labor in the production process made such innovations even more likely because, as workers specialized in one task alone, such as weaving, they, or those managing them, likely were thinking of better ways to do this singular task. And once the mechanical flying shuttle weaver did, in fact, emerge, it put that much more pressure on thread spinners to supply thread more rapidly, hence the pursuit for something like the spinning jenny. This is known as *sequential pressure* in the production process and it remains an important impetus to innovation to this very day. Second, successful inventors could make enormous profits of their own on the basis of such industrial innovations. Again, this is the significance of city-derived merchant culture and the freeing-up of individuals to make their own status in the world. Third, the successful substitution of machine for potentially unreliable, unpredictable, and troublesome human labor at this point becomes a very powerful incentive for the search for ever-more machine-based innovations in the production process on the part of those who control it. In short, for all of these reasons, such innovations have been forthcoming on a continual basis from the Industrial Revolution to the present age of job-killing, labor-saving automated bank teller machines, self-service gasoline pumps, and, indeed, self-checkout machines at the local supermarket.

CENTRALIZING THE PRODUCTION PROCESS

Overall, then, the division of labor within the process of production expanded output dramatically at the same time that it took the skill out of the worker and put it in the production process itself. Now relatively unskilled workers could take part in the production of the final product. At this point, of equal importance is the actual gathering of spinners and weavers of textiles from their rural cottages into central locations under one roof: the textile factory. Factory production allowed for the use of new steam-based machinery as well as better-quality control and overall management of the production process. For cities, it meant that new and quite massive multi-story factories came to dominate their centers. As noted, it also meant that new rapidly expanding and quite densely packed working-class neighborhoods were growing just outside factory gates. Finally, the new factory mode of production meant that specialized industrial districts began to emerge, the growth of which fed on itself. For example, as one cotton textile factory came to locate in a city, another would follow, and then another as each more recent factory could make use of the growing labor pool and expanding raw material supply and retail output networks already created by earlier textile factories. Later factories thus came to accrue what are called **agglomeration economies**, as described in more detail in Chapter 5.

As factory-based production expanded to other sectors of the economy besides textiles, cities became ever-more congested, with densely packed factory and housing structures, some of the latter little more than overpopulated, severely unhealthy hovels. The new coal-based power supplies, including for home heating, also made the air of evolving industrial cities heavily polluted, and to the extent that the sewer and freshwater supply infrastructure could not keep up with city growth, as it most often could not, environmental pollution was a quite general condition. For these reasons, industrializing cities became ever-more densely packed, congested, dirty, and, ultimately, unhealthy for all city people. They also became increasingly dangerous for wealthy city people. Once living in the center of the city close to their factories, for example, wealthy owners increasingly found themselves surrounded by a new class of relatively unskilled, low-paid, somewhat restless laborers living like sardines in their hovels right outside the owners' front doors.

Finally, the last major innovation of industrialism important for cities involved major changes in the *modes and networks of transportation*, as alluded to above. Not only did steam-powered locomotives and ships greatly expand the resource and market hinterland for city-based production, they also greatly accelerated the process of meeting supply and demand needs in the marketplace. This necessitated further innovation – for example, in making productive use of rapidly growing supplies of raw materials and in successfully expanding sales of increasing amounts of final product. Furthermore, the new modes and networks of transportation moved more than just raw materials and goods for sale, they also moved people much faster and cheaper over longer distances, particularly overland. This made it far easier for people to migrate to cities for work or, indeed, to migrate out of cities either for brief bucolic repose in the countryside or for more permanent escape from the growing health and social dangers of the burgeoning industrial city. And, as discussed in Chapter 4, the automobile will eventually provide an even more efficient means to move people in this way.

URBANIZATION AS A CONTINUAL PROCESS OF RURAL-TO-URBAN MIGRATION

The Industrial Revolution was thus a city-based and city-creative process. Cities and their populations began to grow continuously from this period forward. With the social division of labor and growing demand for agricultural products, for example, came innovations in agricultural production which were also mainly labor saving. New methods of production, from better plows, to different planting patterns, to, particularly, the consolidation of many small landholdings into much more commercially efficient large landholdings, meant that fewer jobs were available in the countryside. This was especially the case in the immediate hinterland of cities where demand was rapidly increasing for agricultural produce. At the time, the vast majority of people, even in the first industrializing country of England, lived in rural areas or very small rural villages where they eked out a mostly subsistence living. However, this vast majority of people did not own the land on which they worked. Instead, they rented it from the very few, very large landowners who did. With growing demand for agricultural products from the cities, including wool for textiles, landowners and, significantly, city-based investors in land found that they could make more profit from the sale of such products than they were making from the rent paid by their tenants. They thus began to consolidate their lands behind large enclosures; in the process, their many tenants were, quite literally, thrown off the land.

This activity of enclosure thereby helped to create what looked like a surplus population in the countryside which had little or no opportunity to find gainful employment for survival. The only opportunity for this was to be found in cities, or so it seemed. The new overland modes and networks of transportation in many cases rendered this rural-to-urban migration more rapid, as cities began to fill up with this population recently rendered surplus in rural hinterlands. Some of these new immigrants to the city, particularly lower-paid women and children, found work in the growing factories, of course, but, importantly, not all. This created a situation where it did, in fact, look like a major proportion of this new city population was surplus, over and above the level that could or, to some observers, indeed should be properly kept alive based on the actual and foreseeable rate of output of food products. Of significance, it was the very fact that this population was congregating now in densely packed cities that made it *seem* that this was the case. In other words, the very density of city life magnified the "problem" to observers unused to seeing such densities of population, particularly in such a destitute state, on such small territories. As discussed throughout this book, this is a common perception of the city experience with regard to social issues in general.

THE CHANGING LOCATION AND MORPHOLOGY OF INDUSTRIAL CITIES

All of the changes that took place in the process of industrialization changed the economic efficiency of city location dramatically. Now cities could, and did, develop rapidly in more inland settings, either where raw materials such as coal or iron ore were actually mined or where these materials were put together in the production of iron and steel somewhere in between. Cities also sprang up along railroad lines as **break of bulk sites** for goods from proximate hinterlands to be transported to more distant markets. Finally, cities arose at those places where railway lines converged to form new networks for steam-based locomotive transportation. Similar changes will take place with the introduction of automobile travel and road networks.

The internal social and built morphology of cities also changed with industrialization, as alluded to already. Railroad lines crisscrossed cities, rendering quite intrusive material divisions among neighborhoods, and converged in major central stations, usually in very close proximity to the center of the city. The trains that arrived at these stations brought with them the major raw materials for the factories of the city and also the growing masses of people migrating there. This created a centrality to the railroad stations themselves, both in terms of break of bulk to new city transportation for raw materials and increasing numbers of consumers for products made for sale in their vicinity. The great factories became the new built cathedrals of the industrial city center, congregating in districts of specialized production. And, as mentioned, new working-class housing was built in close proximity to the factories, generally consisting of densely packed multifamily structures of several floors to pay for the higher ground rent of much-sought-after central city land. Such housing grew rapidly as the city-based industrialization process continued, rendering the laboring classes increasingly the majority of city populations, quite centrally located.

Industrial cities also began to expand along the new railway lines. Rail lines, in other words, not only brought people to the cities, they also made it easier for people to move out. Internal to the cities, first horse-drawn carts on rails and then steam and electric streetcars on rails provided the wealthy and middle classes with the means to escape the growing human and environmental congestion of the central city to locate in houses within commuting distance. So-called **streetcar suburbs** of housing for the wealthiest began to appear, spreading city life spatially along these new modes and networks of transportation. Businesses catering to the wealthy began to follow suit.

Finally, the longer-distance railroad system allowed even longer commuting distances, eventually creating a star-shaped expansion of the built and social morphologies of industrial cities along these lines.

ADJUSTING TO INDUSTRIAL CITY LIFE

The new industrial factory-based mode of production represented quite a change in the lives of people in the city, and this change, as discussed in more detail in Chapter 7, eventually came to be associated with city life, in general. Industrialism, for example, entailed a much greater emphasis on mechanical clock time and the quantity, rather than the quality, of what was produced for sale on the market. Traditionally, agricultural labor in the countryside was based on natural day and seasonal timeframes and was thereby quite cyclical. In turn, even the lowest classes had some control over their work process and certainly the time to undertake many other personal tasks in addition to laboring in the agricultural fields. The industrial work process, however, ran by artificial linear time, as laborers clocked into work in the morning, had set times for breaks and for eating, and equally set times to clock out of work at the end of the day. Factory owners, of course, wanted as much work time as they could possibly get out of their laborers; so, often, the working day was set for 12 hours or more, six and sometimes six and a half days a week. Time was money, as it is still said, and the more time laborers were set working, the more profit the factory owner made.

This move to clock time was a difficult one to make for many new city people, leading to high turnover rates for jobs and to resistance among industrial laborers, sometimes even violently so. Similarly, the division of labor in the factories proved to be quite a dreary prospect for workers. Doing one thing all day, day after day, was severely monotonous at best, overwhelmingly mind numbing at worst. And this was particularly the case as workers felt chained to their work area for so many hours a day. In the end, this new mode of factory production led many observers to argue that industrial city life itself was a danger to the mental and physical health of city dwellers. Seemingly, the hustle and bustle of mechanical time pacing, the dense crowds of dull-minded strangers looking for distraction, and the likely groups of entirely disaffected, potentially anti-social people who could not make the transition to industrial life made cities dangerous for everyone. In this context, the only solution seemed to be to get out of cities as best as one could. In short, this growing perception of industrializing cities became an indictment of such cities as cities and not, as it should have been, of the particular way in which industrialism was being undertaken by those who most controlled the process.

Again, there always has been an element of danger associated with city life, either because of the dense packing of different social classes in small territories or because of the rise of new classes of merchants or property-less laborers representing a threat to elite authorities. Largely because the rise of industrialism resulted in a continual growth of city populations, however, it is important to recognize that this has become a more general indictment of city life itself. Such a view is certainly behind the continuing desire of many people to escape to the suburbs and exurbs, as discussed throughout this book. In the United States, particularly, the ideal has long been to either avoid the development of cities altogether or, if that is not possible, to move out of cities as soon as one can move up the income scale. This anti-city sentiment which grew most severe during the Industrial Revolution thus unfortunately still lingers, rendering it that much more difficult to garner much support for fixing the built and social problems that exist in so many cities today.

In any case, by the end of the nineteenth century, the division of labor in the factory as well as the new energy sources and transportation modes and networks of industrialism had spread to most major sectors of the economies of industrializing countries. Steam engines were continuously improved and put to use in factories, as well as on farms. Innovations in the smelting of steel drastically reduced its price, rendering steel an ideal material to make engines and other machines stronger, more reliable, and even cheaper than previous machines made of wood and iron. Cheap and durable steel also eventually became a significant factor in the built morphology of cities as it made much higher building heights possible both in terms of engineering and cost. And new steam-powered elevators and innovations in communication, such as the telegraph and, particularly, the telephone, which came on the market in the 1870s, made such higher-rise buildings conducive for day-to-day business transactions.

Finally, as the new industrial innovations and factory-based production processes spread throughout economies, more commodities flooded commercial markets at much lower prices. For reasons already discussed, such lower prices brought more and more people into these markets, thereby beginning a new phase of industrial development commonly known as **mass production and mass consumption**. The key is that the two phenomena essentially are linked in this manner, although this will not come to be entirely understood by most people until the innovations in the organization of industrial production linked to the name of an American, Henry Ford, in the early part of the twentieth century, as detailed in Chapter 4.

GROWING CITIES OF MASS PRODUCTION AND CONSUMPTION

The spread of new industrial techniques to more and more sectors of the economy was hastened considerably by further innovations in the three areas of production treated earlier in this chapter: *sources of energy*, *the organization and technology of production*, and *transportation modes and networks*. In terms of energy, the turn of the twentieth century was a time of increasing use of electricity and petroleum. Once used mostly for lighting, which had enormous impact itself upon city night life and, indeed, in terms of the ability to keep factories producing around the clock, electricity increasingly began to be used in the production process itself. Electricity as a source of energy turned out to be quite an improvement over coal-based steam power because electricity is a much more "portable" form of energy. It took much coal to keep larger steam engines running, which meant that steam-powered factories had to be located near major coal deposits as the cost of digging and, particularly, transporting such amounts of coal was prohibitive. Electric energy, on the other hand, can be created at specific sites, usually on waterways and later at coal-fired plants, and then be conveyed relatively cheaply across increasing distances. This eventually widened considerably the locational possibilities for industrial factories.

Electricity was also portable in terms of its ability to power large factories without itself taking very much factory space. Steam-powered factories had to be powered from large-scale coal-lit engines within, or very near, the actual production spaces of whatever was being produced. This necessitated both the building of such engines and factory space, as well as the building of compact factories as a result of the lack of portability of the energy produced. Steam-powered factories, to be efficient in energy use, thus tended to be expansive and multistoried, creating a city skyline of higher-rise, smoke-bellowing, fortress-like built structures. The portability of electric energy, however, changed this situation in two main ways: increasingly, cities became dotted with electricity-conveying substations redirecting electric current to where it was needed; and, more importantly, electric energy rendered it more and more possible to produce whatever was being produced all on a single-level factory floor. As we will see, this latter possibility became increasingly important as factories began to move out of overly congested industrial cities. Here, it is significant to note that this new possibility for horizontal production space rendered it much easier for owners and managers to oversee the entire production process, which, in turn, rendered it much easier to control the every movement of factory workers throughout the entire working day.

Electricity also made it possible to create ever-smaller motors to power the actual tools that workers used at their work stations. This further mechanized and, in most instances, accelerated the production process, resulting in more products on the market for sale. It also led to spin-off industries producing and repairing such machine tools, adding to the industrial base of city economies, in general.

Finally, there was an increasing use of petroleum as an energy source which has many of the same properties and uses as electricity. However, the most significant impact of the increasing use of petroleum was twofold. First, the conversion of shipping from coal-based steam to petroleum power increased the hold capacity of ships, now no longer needing to carry so much coal as fuel, and their speed, as petroleum power was much more efficient. This was a boon for long-distance trade particularly. Second, the use of petroleum became enormously important with the increasing perfection of the internal combustion engine as a means to power automobiles and trucks.

REORGANIZING THE CITIES OF THE IMPERIALIZED WORLD

When Spain and Portugal set out attempting to find a sea route to the East, they were not looking to conquer lands and people.

Rather, they were essentially after three things: trading goods, trading partners, and, if nothing else, gold and silver. Spanish and Portuguese eyes were focused on those things considered most important for success in early Renaissance power struggles in Europe, particularly between the Iberian monarchies and the merchant city-states of Italy and Flanders. That they would also by-pass infidel Muslim traders of the Middle East was a bonus, as mentioned in Chapter 2, but still important in the minds of many Iberian explorers overwhelmingly influenced by the culture of Middle Ages Christendom. Finally, that the Spanish and Portuguese sailed in the name of the Crown and not for private merchant capital simply solidified this cultural view.

In the East, of course, the Portuguese did find the traders and trading goods that they were after. Africa proved a more difficult region to deal with. The great capital and trading cities of Africa at this time were located inland and difficult to make out from coastal waters; when they did become known, attempting to penetrate that far inland proved deadly for Europeans unable to withstand tropical diseases. The Spanish found much the same thing in their new world of the Americas; but unlike Africa, it was disease carried by Europeans that was the scourge of the indigenous population. This, as well as differences in war strategy and technology and the hatred of already conquered indigenous societies for the Aztec and Inca empire builders, allowed Spain to conquer the great civilizations of the Americas quite rapidly at the same time as Portugal took the coasts of Brazil from indigenous hunters and gatherers. Vast territory and large numbers of indigenous people thus came under the imperial control of European conquerors who then had to protect their conquest from European competitors, particularly the ever-more powerful Dutch and English. To do this, the Spanish built their first colonial cities on top of those of the earlier imperial powers, such as Mexico City, Lima, Quito, and Cuzco, as a means for internal administration and control.

The reason that early European explorers did not want to conquer peoples and territory was that, if they did, they would have to protect these lands from other Europeans and this would be a severe drain on any profits made from trade. That the Spanish and Portuguese did come to rule over the Americas thus made it necessary to find, as rapidly as possible, something of value to pay for the protection of this conquest and, hopefully, turn some profit for their crowns back home. Luckily for the Spanish, they were able to find important sources of gold and, particularly, silver to send back to Europe, eventually by the ton. Yet, at the same time, other sources of value were being created on the islands of the Caribbean and on the coast of Portuguese Brazil. This was the production of crops, particularly Brazilwood for dye and sugar, which could be sold for profit back home in Europe. Since such **cash crops** were produced simply for trade and profit, they were produced in the most efficient way possible on the most fertile land divided into very large plantations worked by laborers. It was the need for a steady supply of labor for these plantations that was the source of the rapidly growing slave trade from Africa, with all the ramifications this was to have for both Africa and the Americas.

Here, however, the emphasis is on this organization of internal American production for external reasons. From the point of view of profit, it simply did not matter that all of the most fertile land was taken up with sugar (or cotton or tobacco or tea) production. In fact, this was the most efficient way to produce these sorts of cash crops even if food had then to be imported to the area of production. Nor did it matter that slaves might not be the best, most reliable laborers because of their severe lack of interest in the result. Slaves were just as much commodities to be bought and sold as the sugar they produced. The point is that this early organization of cash crops for export in the Americas provided the model for the reorganization of the internal production patterns of European-held territory that would soon become much more common worldwide in the later period of industrialization.

Indeed, with industrialism came the increased demand for two things: raw materials as inputs for the factories and consumer markets for their increasing output. In this period, then, Europeans, now including the Dutch, English, and French, particularly, literally scanned the world for territories to produce raw materials exclusively for export and with indigenous populations to which the final product, produced in Europe, could be sold. During this period, then, the conquest of territory and peoples was very much on the European agenda, and by 1914 it is estimated that Europeans had come to control as much as 85 percent of the Earth's land territory and all the natural resources and peoples that this territory included.

CITIES IN THE NEW INTERNATIONAL DIVISION OF LABOR

European-led industrial imperialism thereby created an **international division of labor** with Europe at its center. Conquered peripheral countries and peoples in the Americas, Africa, the Middle East, and Asia supplied Europeans with raw materials, particularly raw cotton and peanut oil for industrial lubrication, for their factories and consumer goods (such as sugar, tea, coffee, and tobacco), for final consumption. This imperial relationship was thus a win–win situation for Europe. As more and more raw materials were forcibly procured from imperial possessions, more and more factories arose in European cities and more and more final commodities were produced. This lowered the prices for such products in the European market, thereby extending this market as discussed above. Having captive markets for such goods in their imperial possessions, however, also allowed for the ever-growing amount of such industrial products to be sold at a profit before prices dropped below the costs of production as a result of an actual glut of such products in the home market. In short, this international division of labor greatly accelerated industrialization in Europe, resulting in ever-growing numbers of cities and city people. Indeed, many other secondary industrial cities grew in addition to the major merchant cities of the earlier era. This marked the beginnings of a **nested hierarchy of cities** within Europe of very large cities, secondary-sized cities, tertiary cities, etc., which spread urbanization quite evenly in most countries.

In the imperialized non-European peripheral world, however, the opposite was occurring. Internal production was set to producing raw materials and European-bound consumer goods on the best, most accessible lands for strictly export, cash-crop purposes. Most often this led to severe monocultural production patterns in which one or two crops took up all the best agricultural land in large-scale plantation-like forms of production. Indigenous people were forced onto ever-more marginal land for the production of their subsistence crops. Most often, they were also forced, either outright or by various means of taxation, to work on the plantations producing the main cash crops for export.

This new imperial structure of internal development resulted in three main outcomes. First, opportunity for material advancement for indigenous people in rural areas was very limited, as all the best lands were controlled by European imperial authorities or their commercial representatives. Wages, if they were extended at all, for plantation labor usually were just enough to pay the taxes imposed by these same authorities. Second, internal transportation modes and networks were either restructured or constructed anew with a focus on gathering the cash crops or other raw material output from the place of production to coastal port cities for export to Europe. This was usually done by railroads along fixed tracks converging on these ports. In this respect, there was no real incentive to create a crisscrossing network of tracks for internal regional development linkages.

Clearly linked to these first two, the third result of this imperially imposed structure of development was a much strengthened water and mostly coastal focus for cities in the Americas, Africa, the Middle East, and Asia. Not only that, but the singular focus on producing the one or two cash crops or other commodities for export, as well as the nature of transportation infrastructure, rendered it much more efficient to focus on one or, at most, two main port cities to handle the bulk of what was to be exported. These ports thus became tips of a funnel-like process to get the goods out of the country as quickly and efficiently as possible. Indeed, this funneling process also included indigenous people from the rural hinterland seeking opportunities for a better life that could not be found in the countryside. In this way, the urban structure of imperialized countries came to consist of one or two major port cities that continued to grow in population and economic centrality with hinterlands of cash crop and other raw material production which included only small villages. This is known as a **primate city** structure, which has fed on itself even after the formal independence of European imperial possessions. Bequeathed with the stunted development pattern of raw material production left by former imperial powers, many countries today continue to exhibit this urban structure. The fact that population growth rates increased rapidly in the post-World War II period and that rural opportunities did not, as discussed in Chapter 4, simply fueled this pattern as there was nowhere else to go for opportunity but the one or two major primate cities. In short, this pattern of over-urbanization in former imperialized countries and all the social and environmental problems such over-urbanization includes is a direct legacy of the international division of labor constructed by Europeans in the industrial age.

Box 3 Tenotchtitlan: The capital city of the Aztecs

After the fall of Teotihuacan, other major cities eventually evolved in the Basin of Mexico. Tenotchtitlan, for example, was founded in 1325 CE and existed as such until its capture by the Spanish in 1521 CE and rechristened Mexico City. Tenotchtitlan was founded by a Nahua people, the Mexica, at the west end of a large shallow lake, Texcoco. The city was joined to the shore by large built causeways which allowed access for people and goods. The leaders of this city-state eventually formed an alliance with those of two other Nahua city-states, Texcoco and Tlacopan, to form the Mexica-dominated Aztec Empire that came to rule over most of the territory of modern day Mexico. Tenotchtitlan thereby became the capital city of the empire which commanded resources from a more extensive hinterland. Because of this, Tenotchtitlan grew to what some have estimated to be a population of as much as 200,000 before it was conquered. This was more than double the size of Seville, the largest city in Spain in the late 1400s CE.

Indeed, when the Spanish explorers first caught sight of Tenotchtitlan they could hardly believe their eyes. The massive, light-toned, almost glittering monumental architecture of the Aztec city literally rose straight out of Lake Texcoco as if it were on a cloud. The great joining causeways hustled and bustled with goods and people; once inside the city, the Spanish were profoundly awestruck by the planned linearity of its main streets as well as their overall cleanliness and spatial expansiveness. The central marketplace was larger than anything they had ever seen in Europe as perhaps as many as 60,000 traders plied their wares. The Spanish even remarked on the efficiency by which the Aztecs disposed of their human waste in the city, ensuring that it was regularly taken from nearby waterways to be used as fertilizer for local agricultural production. Overall, the Spanish could not help but be astonished by the splendor of Tenotchtitlan. To many of them, the Aztec capital and its designed waterways brought to mind the glories of Venice, if on a much grander scale. Not only was it the largest city they had ever seen, but it was also quite a contrast from the densely cramped, waste-filled squalor of most European cities and towns of the time.

Of course, this awe on the part of the Europeans did not make them hesitate to conquer the city and the empire that it ruled by the most violent force they could muster. Indeed, the Spanish conquest of Tenotchtitlan in 1521 CE provides a very clear example of how the global spread of Westerners after 1492 CE has had a most powerful impact upon cities and city people everywhere. The European conquerors took complete control of the most central sacred parts of the city, and quite literally dismantled much of the monumental architecture of the Aztecs to provide building materials for new European-fortified residences and Catholic cathedrals located there. Catholic chapels were even constructed on top of the most massive pyramids to signal the power of the new masters of what was now to be known as Mexico City. In short, the great Aztec city of Tenotchtitlan was materially and socially rearranged to reflect the power of new European masters. In this respect, Mexico City today can be considered a prime example of a dualistically layered colonial city as a result of Western imperialism.

FURTHER READING

A very good overview of world city trading networks before the rise and overwhelming global dominance of the West is Janet L. Abu-Lughod's *Before European Hegemony: The World System AD 1250–1350* (New York, NY: Oxford University Press, 1989). A good general and readable introduction to city history is John Reader's *Cities* (New York: Atlantic Monthly Press, 2004). A classic history of cities, particularly in the West, is Lewis Mumford's *The City in History: Its Origins, Its Transformations, and Its Prospects* (New York, NY: Harcourt, Brace, 1961). A more recent overview of this history focusing more on the built environment can be found in James Vance's *The Continuing City: Urban Morphology in Western Civilization* (Baltimore, MD: Johns Hopkins University Press, 1990). Finally, a very comprehensive overview of the spread of Western influence and dominance across the planet from 1492 on can be found in Immanuel Wallerstein's multivolume account of *The Modern World System* (New York, NY, and Berkeley, CA: Free Press and University of California Press, 1974–2011).

4

FROM INDUSTRIAL TO POST-INDUSTRIAL CITIES

Industrialization thus brought great changes to cities and to the lives of city people. Large-scale factories and working-class housing filled up the built environment of city centers and the new, relatively unskilled, working class became a major part of the population. Such changes were rapid and very dramatic as predominantly rural agrarian societies became largely urban industrial societies almost overnight. With very few controls over the market processes that created such dramatic change, new industrial cities grew ever-more congested, with often substandard built structures and more and more people packed together on smaller and smaller territories. As new modes of transportation arose, even the streets of industrial cities became congested with increasing traffic and people attempting to move about. Finally, the environmental quality of city life deteriorated dramatically as new factories belched smoke from coal-fired furnaces and inadequate infrastructures and services were overwhelmed with vastly increased amounts of human-generated waste.

It is no wonder then that many city people of the time began to think of better ways to organize city life, as will be discussed in some detail in Chapter 9. It is also no wonder that, increasingly, those city people who could afford to began to move out of the central areas of the city toward the periphery. Indeed, as the industrial process matured into the period of mass production and mass consumption, increasing numbers of even more middle-class city people began to follow the outward migration of the wealthy to the suburbs. Because of innovations in transportation mode and network, particularly the internal combustion engine and paved streets and highways, this outward migration was more evenly distributed across formerly rural territory. From distinctly circumscribed streetcar suburbs, in other words, during this period much of the suburban territory between rail lines now began being filled up in a process known as **suburban infill**. Before discussing this process in more detail, however, it is first necessary to outline the specific process by which mass production created this possibility for the out-migration of increasing numbers of city people via the creation of mass consumption.

HENRY FORD AND THE MASS PRODUCTION OF AUTOMOBILES

Chapter 3 discussed the sources and implications of the major innovations that took place in the organization and technology of industrial production by the early twentieth century. The factory division of labor, for example, was increasingly refined and mechanized and became standard in sector after sector of major industrial economies. But the most important innovations in the organization of city-based mass production in the twentieth century have come to be associated with the name of Henry Ford, an American entrepreneur very early involved in the emerging automobile industry. Indeed, many urban scholars today associate Ford's name with an entire epoch of economic growth after World War II as a result of the greater social impact of these innovations, as discussed later in this chapter.

Ford basically fine-tuned the factory division of labor by bringing the work to the worker at each station of the production process by means of an automated conveyance system. If an individual's job was to install windshields all day in the automobile factory, for example, partially constructed automobile bodies would be sent to them at regular, strictly timed intervals in order to perform their task before the partially constructed automobile went off to another worker to perform their respective task, say, of polishing the windshield. In other words, individual autoworkers would not any longer be roaming around the factory to each automobile being constructed by a work team. Instead, autoworkers would now stay in one place all day, performing their tasks at regular intervals depending upon the speed of the automated conveyance system. This is such a common practice in today's assembly line manufacturing process that it seems hardly innovative. Yet, at the time, it was highly innovative and greatly accelerated the production of automobiles, lowering the price of each dramatically for increasingly mass consumption.

Besides automated conveyance, other innovations first successfully applied at the Ford Motor Company would come to be adopted in industrial manufacturing, in general, as well. To be entirely successful, for example, the acceleration of the assembly process had to rely on increasingly standardized and interchangeable parts. While most automobiles manufactured at other companies at the time were handcrafted with special luxurious adornments and options that catered to the individualized consumer tastes of the wealthy, Ford's automobiles were simplified, standardized, and came with little adornment. For many years, they even came out on the retail market in only one color, black, because Ford's plants produced them so rapidly that other colors would not dry fast enough for the last stage of production.

Planting the worker in one place at their particular stage of production also proved quite efficient. Not allowed to roam the factory, workers suffered far fewer injuries in the factory and, as a result, there were far fewer interruptions in the production process. At the same time, it was easier for owners and managers to keep track of how each individual was performing their specific task. This is particularly the case because in an automated assembly line, if one worker is not keeping up with their task in a time appropriate manner, it slows the whole process down. This led to whole series of time management studies to determine quite precisely the appropriate length of time a worker should perform their specific duty in order to be most efficient in terms of the assembly process as a whole. The most famous of such studies were produced by Frederick Taylor, which is why **Taylorism** is usually associated with what has come to be known as **Fordism**, or the legacy of Ford's innovations in the organization of mass production.

This new, automated assembly-line organization of production was most efficient in terms of output and the price of automobiles.

Significantly for the future of cities, it was also best achieved in more expansive horizontally constructed production plants. For this reason, such plants began to be located outside of city centers where land was more widely available and much cheaper. This type of mass production thus very much helped to stretch the territorial extent of cities outward as plants were located increasingly on suburban (and eventually) exurban **greenfield sites**.

FROM FORD TO FORDISM

Nevertheless, Ford soon found that the new organization of production also resulted in quite a bit of unrest among autoworkers. Another reason to seek more suburban locations for new mass-producing factories was the very restlessness and beginning union organizing and activism of city workers, not just those employed by Ford. Such activism, such as strikes, work slowdowns, and sometimes even industrial sabotage spread rapidly among workers in the city, densely packed as they were around their workplaces. Many a new Ford worker also quit very soon after taking a job at the new-fangled assembly plants. The monotonous and continuous assembly work and the lack of free mobility took a real toll on the minds and bodies of the mass-producing workforce. Now, even more so than when the factory division of labor first arose, the worker felt little more than a trained monkey performing menial tasks at fixed, and strictly enforced, intervals of time, with little personal input or freedom of maneuver throughout the entire work day. Even difficult, time consuming, and less lucrative farm labor seemed, to most workers, better than this.

Recognizing this growing problem of labor unrest and turnover, Ford instituted a policy which, at the time, seemed to be preposterous from a business point of view. He raised the wages of his workers to more or less double what they used to be and what was the norm in all other automobile factories at the time. In exchange for this much higher wage, Ford instituted a code of conduct for his workers, including how they conducted their private lives outside the factory. Of importance, this policy largely worked, as labor turnover rates dropped rapidly in Ford factories, resulting in much fewer delays and actual interruptions in the production process. But this new higher-wage strategy also produced another result that benefitted Ford. As workers took home more pay they increasingly found themselves able to afford Ford automobiles, the price of which continued to drop. Given Ford's further experiments in preferential pricing for Ford employees, as well as installment credit plans for the purchase of his automobiles, it is clear that Ford's seemingly preposterous wage scheme actually accelerated the creation of a mass consumer base for his mass-produced product.

Indeed, this last paragraph holds the key to why Ford's name is associated not just with innovations in assembly-line mass production, but also with wider social policies in the regulation of production and consumption, in general. His attention to satisfying, or at least assuaging, the concerns of his labor force at the same time as continuing to turn healthy profits was a major aspect of the post-World War II economic boom, particularly in the United States. That the Ford Motor Company was allowed to essentially corner the market for automobiles in these early decades of the twentieth century, particularly during and after World War I, as long as it provided jobs, is also part of the general picture of the post-World War II economic boom years of compromise among big business, big labor, and big government. In short, these economic and social aspects of Ford's business practices are what will turn our attention from Ford the man to what has been called Fordism as a mode of production and consumption, as discussed below.

AUTOMOBILITY AND THE CITY

As noted in Chapter 3, the internal combustion engine was perfected in this period and this had enormous consequences for cities. With the generalization of Ford's innovations, automobile ownership in the United States soared as prices dropped and installment credit allowed more and more people to own cars. It has been estimated, for example, that from about 8,000 automobiles in use in the entire country in 1900 there were as many as 500,000 by 1910 and, incredibly, a total of 27 million in 1930. This was, indeed, the beginning of the age of the automobile. With ever-rising levels of automobile ownership and demand came, in turn, the need for more and better roads in addition to more and more factories in production. City streets became quite congested with this new mode of personalized transportation crowding in on traditional animal- and electric-powered modes and networks of transportation, both private and public. This road congestion made doing business in the city increasingly difficult in terms of both procuring raw materials or wholesale supplies from outside the city and in getting final goods to equally extra-local markets. Along with the new possibilities for single-floor production, more and more factories thus came to be located outside city centers for this reason, as well as that of much cheaper land prices. The same is true for wholesale warehouses in an attempt to bypass the growing gridlock of city traffic, at least for trucks coming in with products for city markets.

Of most importance is that this new mode of transportation, both automobiles and trucks, ran on a network that was not fixed and linear in nature like railroads and streetcar lines. Essentially, roads can be constructed just about anywhere with as many curves and angles and criss-crossings as can be conceived, or at least as are feasible engineering-wise. While railroads and, particularly, streetcars already had allowed many city people and businesses to move increasingly out of the city center, the automobile and trucking age accelerated this process quite considerably. Local taxpayer money began being put into road construction given demand for such among the increasing mass of automobile and truck owners. As early as 1916, even federal funds were being designated for this purpose in the United States. By 1921, the focus of the federal government already had shifted to a long-distance road network. In the end, as more and more roads were constructed, more and more people and businesses found it possible, even desirable, to move out of increasingly congested and high-cost city centers and at greater distances. The automobile thus began, and greatly accelerated, what has come to be known as **urban sprawl**.

CITY LOCATION AND MORPHOLOGY IN THE AUTOMOBILE AGE

The new transportation mode and network of automobiles and trucks considerably changed the locational efficiencies of firms and people. Suburbanizing factories no longer had to be located along fixed rail lines and neither did people. From a star-like pattern of the built environment along these rail lines, automobiles and trucks thus allowed for considerable suburban infill, as noted earlier.

Similarly, the locational efficiency of towns and cities changed to the extent that networks of roads and, eventually, highways increasingly linked them to larger cities and their markets. Unlike in poorer countries with their primate cities, this increasing linkage of towns and cities in richer countries ensured the continuing development of a nested hierarchy of urban places, from the very big cities to the not so big to the middling big to the smaller and on to the smallest town. This is a hierarchy not just in population size but also in economic diversity and extent.

Increased road building and ever-rising automobile and truck ownership also began to stretch cities as ever-larger swathes of formerly rural territory outside city limits was taken up in industrial plant and housing. Importantly, this was a self-propelling process. Automobile production was increasingly undertaken by large-scale corporations such as Ford and General Motors, which were allowed to become so big on the basis of the federally funded war effort and its demand for armored and other vehicles, as alluded to above. "Bigness" allowed more efficient production as such corporations could accrue **economies of scale** the larger they grew. More vehicles being produced more efficiently also increased demand for such inputs as rubber, steel, and petroleum, and a few large corporations began to dominate these industries as well. After World War I, as production restructured for the consumer market, economies of scale accruing to these large corporations allowed them to continue to dominate their markets. Such economies of production also meant that prices were generally dropping, as manifested again in the extremely rapid increase in automobile and truck ownership.

The large corporations involved in automobile production thus had a major interest in the construction of roads. Because of their economic importance and financial resources, they also had the political clout to get this done. In turn, new federal and local taxes on petroleum helped fund the construction of roads and, particularly, highways. In short, the demand for road construction came from the rapidly increasing numbers of automobile owners and from those who produced them and their suppliers, as well as, finally, from governments interested in keeping all of these groups satisfied for political reasons.

So automobility greatly stretched cities in the early decades of the twentieth century and continues to stretch cities today. Wherever roads can be built, businesses and people will come to be located. This still explains much of the differences in the morphology of East Coast and Midwestern cities and those of the West and, most recently, the South in the United States. In Boston and Philadelphia, for example, one can still see the central focus of the pre-industrial city in the dense built environment around their ports. Similarly, the wider star-like spread of the built environment of Chicago along original rail lines can still be easily discerned. In automobile-era Los Angeles, however, the spread of the built environment really has been more like lava flowing out over larger and larger swathes of territory in a relatively even pattern. The same is true in the Tampa Bay metropolitan region from which the present author hails.

REMODELING THE AUTOMOBILE CITY

Finally, as more and more factories moved to the suburbs to take better advantage of the new technology and organization of mass production, the built and social morphology of city centers also began to change. First, as firms became large-scale conglomerates like Ford, they needed more **white collar workers** to attend to things such as management, accounting, advertising, legal services, sales, and research and development. These are essentially office jobs and are best done together in close proximity for a coordinated focus on the bottom line of profit for the entire corporation. Indeed, many still argue that, even in the current age of internet communications, these types of jobs necessitate face-to-face contact among colleagues or even competitors to be performed most efficiently. Of importance, save for some managers, these jobs do not have to be done right at the factory production site. This is particularly the case with regular innovations in communication, from the telegraph/telephone through to the FAX/Internet age. Increasingly, such office jobs were located in center city locations, both as show-brand headquarters for major firms and as a means of keeping up with the latest moves of close competitors already located there.

Center city location, particularly in the larger cities, thus had a functional role as it continued to retain a certain prestige factor. Now office space, not factory space, came increasingly to dominate these city centers in taller and taller buildings made possible by cheaper steel and steel-construction techniques. In fact, elevation became a necessity for all businesses and residences in city centers because land was so expensive. This remains the case today. Building up ever higher on small pieces of city territory spreads the cost of this land among all the tenants of the office building, retail structure, or apartment/condominium building. And the reason land is so expensive in the center of cities is because there continues to be great demand for it. Again, this is a self-propelling process. As more and more people demand central city space for their business or housing, the cost of this land goes up and up in what has been called a **bid-rent curve** fashion. Importantly, this means that only certain types of business and office work will be able to afford such costs, no matter how high they build the building.

The new central business districts (CBDs) thus became dominated by office buildings, including corporate headquarters, independent producer services firms, banks and other financial corporations, as well as high-end retail, dining, and entertainment establishments which catered to the overwhelmingly white collar workforce. And, again, this remains the case today. Unfortunately, these new CBDs were literally surrounded by increasingly abandoned, soon to be derelict, industrial factories and warehouses, as well as still densely packed working-class neighborhoods, the population of which no longer had the right skills to fill the most prevalent jobs available in central cities. Thus began to emerge what has been called a **skills mismatch** between jobs in contemporary CBDs and the surrounding **inner-city** population. And this mismatch has only increased in magnitude in the post-World War II era, as discussed below.

CITIES AND THE WAR EFFORT

Like World War I (1914–1918), World War II (1939–1945) had a devastating effect on Europe, and European cities, as well as on many European imperial possessions and their cities around the world. Also like World War I, the United States emerged again as a global power, after years of relative isolation dealing with the Great Depression of the previous decades. Because the war was fought elsewhere, the economic and social infrastructure of the United States remained more or less intact. In fact, massive federal expenditures for the war effort in Europe succeeded in bringing the U.S. economy out of the Depression by rapidly increasing demand for everything from textiles, to automobiles and trucks, to standardized food, to heavy and light armaments, and so on. This led, in turn, to a vast increase in demand for raw materials and semi-processed goods such as iron and steel, as well as for labor. The war, in effect, got the U.S. economy working again and this brought economic life back to industrial cities, as well as to those along transportation routes and junctions.

But notice how this new economic life was contingent almost entirely upon government spending. Just like World War I, the new war effort directly, and extremely rapidly, increased demand for materials and industrial products. It also was putting money directly in the pockets of increasing numbers of workers in the cities who then increased their spending for things such as groceries, clothes, furnishings, and other consumer products. Grocers and clothing, furniture, and other consumer goods manufacturers then increased their own demand for raw material inputs, labor, and transportation and storage facilities. Firms supplying such raw materials and transportation services then increased their demand for input materials and labor, and so on and so on. Government money spent on final war supplies thus reverberated throughout the economy, multiplying the effect of each dollar spent to many more dollars spent in turn.

This **multiplier effect** of government spending originally was emphasized by the English economist John Maynard Keynes as a means of pulling industrial economies out of the Great Depression before the war. Against common economic wisdom at the time, Keynes argued that the government needs to intervene in this way in depressionary situations in order to increase aggregate demand and thereby restore investor confidence even if this means the government might spend more than it takes in as tax revenue. Keynes's ideas were behind President Roosevelt's New Deal policies of the 1930s, as well as, more recently, President Obama's government-funded stimulus policies to combat the current and deepest recession in the United States since that time. Of importance, budget-busting government demand for the war effort indeed pulled the U.S. economy out of the Depression; equally important, government authorities had both the experience of the previous war and Keynes's ideas to begin to reckon with the post-war period.

Indeed, this last assertion is a key to understanding what is considered to be the extended post-World War II boom in the U.S. economy.

U.S. government authorities began to worry even in the midst of the war years that once the hostilities ceased and government demand dried up, the U.S. economy would very quickly fall back into a depression. The thought was, after a brief period of exuberant consumer spending of pent-up savings from unspent military wages and overall war time rationing like that of the so-called "roaring 1920s," aggregate demand would fall again to depressionary levels. To avoid this, the new Keynesian-tinged economic belief was that the government would have to continue its major new role in the economy by maintaining significant levels of expenditure and otherwise playing a larger role in managing the national economy as a whole.

POST-WAR FORDISM: COMBINING FORD AND KEYNES

Two of the most significant areas of major government spending for cities in the post-war period were in housing and highway construction. In terms of the first, most city people, other than those with significant wealth, still lived in multifamily rental units located in inner-city areas near the CBD. The government-sponsored GI Bill of Rights signed into law in 1944 by President Roosevelt would soon change this. This law provided for mortgage loan guarantees backed by the much empowered Veterans Administration (VA) as a means of rewarding returning veterans for their service abroad. In addition to the provisions for education and retraining, this housing initiative of the GI Bill was believed to be necessary to smooth the transition to civilian life by making property ownership much more widely available. Property ownership, it was thought, was a good means to promote social stability in this time of transition as new property owners would now have a larger stake in the success of the economic system as a whole. The housing provision essentially made it possible for veterans to get mortgages with no down payment. This allowed returning veterans to avoid what is still a major barrier to homeownership. And it proved highly successful: between 1944 and 1952 the Veterans Administration backed nearly 2.4 million home loans under this program.

The impact upon U.S. cities of this government-backed housing assistance was almost immediate. There was simply no space in central cities for the construction of new owner-occupied housing in such numbers and what space did exist was much too expensive to build on. The only alternative was the suburbs and other unoccupied territory on the outskirts of cities. And so the 1950s and 1960s was a period of vast suburbanization of the middle and even lower-middle classes in the United States. One of the first new suburbs was Levittown outside of New York City on Long Island, originally constructed by Levitt & Sons, Inc. between 1947 and 1951. Of importance is that new suburbs such as Levittown came to be constructed using similar production methods as in mass-producing industries. To keep costs down, houses were assembled from mostly prefabricated parts, very rapidly, with no frills. This new mass production of housing thus helped to accelerate the growing mass consumption of housing as prices for each house dropped considerably.

In time there came to be many a replication of Levittown on the periphery of every major U.S. city. Besides the new process of mass-producing housing, there are other aspects of this suburbanization trend that need emphasis. First, house construction, even with prefabricated parts, is a **labor-intensive** industry that requires many raw and semi-processed inputs from sources that also employ a lot of workers. Rapidly increasing employment in construction thus puts a lot of money into circulation, creating more and more jobs as a result of the multiplier effect, as discussed above. Indeed, this is one of the reasons that the current global recession begun in 2008 is so deep: it affected the construction industry most dramatically.

Second, owning a house means that there are many things that need to be bought to fill it up, from appliances such as washers and dryers and refrigerators, to furnishings for each room, to lawnmowers for the new lawn. The point is that some of these things are no longer held in common as in most multifamily housing, but now must be procured by each new owner-occupier. Owner-occupied suburban development thus led to a rapid and continuing increase in demand for such commodities, thereby facilitating continuing development in the industries which produced them and their respective suppliers.

Third, because of continuing racial and ethnic discrimination during the war, the new VA mortgage loan guarantees were overwhelmingly taken advantage of by white veterans. This meant, of course, that most new suburban communities such as Levittown were overwhelmingly populated by whites as African-Americans and other minorities continued to be located in the inner-city ring around the CBD. Indeed, Levittown, from its very beginning, had a quite explicit restrictive covenant which stated that only non-Jewish whites could buy into the development. Similarly, it was mostly the better-off middle-class whites who were able to move to the suburbs and afford to outfit a new home, not lower-class white veterans still trapped by poverty in central cities. This racial and class discrimination was even more the case with the other federally funded housing initiative, which became known as the Federal Housing Authority, which had even more strict requirements for mortgage assistance. This increasing spatial separation between city and suburb based on race and class continues to have enormous implications for city life in the United States.

Finally, these blossoming new middle-class suburbs were essentially **bedroom communities** – that is, people lived there in their new owner-occupied houses, but they worked elsewhere, still largely in more central locations in the city. In fact, this home and work life spatial pattern had a significant gender aspect to it. Men went off to work elsewhere while suburban women were increasingly privatized within isolated houses surrounded by lots of things in the house, but no businesses or truly public amenities and spaces outside the home. Moreover, social networks were difficult to create aside from those involving nearest neighbors who may, or may not, turn out to be that friendly. This isolated suburban living situation of women simply reinforced traditional stereotypes of women as only really fit for the nurture of men, families, and hearth and home.

The other aspect of such bedroom communities important for cities is that they essentially necessitate the building of ever-more roads to and from the city, as well as the ever-increasing use of the personal automobile and commercial truck. This was a boon to automobile and truck manufacturers and the petroleum, rubber, and steel industries. And those who controlled such industries, of course, knew it. The highway lobby consisting of representatives of these industries came together for this very reason: to involve the federal government even more in the suburbanization process.

HIGHWAYS THROUGH AND OUT OF THE CITY

The second major way in which the U.S. federal government remained a prominent actor in the national economy after the war was the maintenance and creation of infrastructure, particularly a new interstate highway system. The Federal-Aid Highway Act signed into law by President Eisenhower in 1956 turned out to be a most powerful impetus of economic development. Sold largely as a defense-related initiative to allow more efficient evacuation of cities under nuclear bomb threat, as well as better mobility of major missiles from site to site, this act earmarked as much as $1.1 billion to construct a network of over 40,000 miles of limited-access highways. The goal, which was attained, was to link directly as much as 90 percent of all U.S. cities with populations exceeding 50,000. With no little overstatement, President Eisenhower proclaimed it at the time the "greatest public-works program in the history of the world."

Given previous discussion, the ultimate effects of this massive highway building project on cities should be obvious. Much employment was generated in the labor- and materials-intensive construction process, of course, which stimulated the overall economy. But this new interstate, interurban highway system rendered it that much easier to conduct city-related business much outside of city centers where land and labor were generally much cheaper. It also facilitated the movement of people farther outside the city. Initially, the pattern of intercity highways was to create a wheel-like highway beltway around cities with spokes of roads leading to the city itself. This pattern created junctions around the spokes where businesses and residences could be located with easy access to the city. As more and more did so, suburban business and retail districts began to emerge, which became the source of so-called **edge cities** as more and more people commuted to work and shopping *within* such suburban locations themselves, rather than to traditional city centers. Indeed, as these edge cities grew in size and diversity of employment, including white collar office jobs, they spawned their own "suburbs" in a literally endless sprawl of urbanized bits farther and farther away from original city centers.

THE COMPROMISE AMONG BIG GOVERNMENT, BIG BUSINESS, AND BIG LABOR

The U.S. federal government thus continued to play a major and quite active role in the U.S. economy even after the war. But, importantly, this continuing presence of the government in the economy included more than directly maintaining high levels of aggregate demand. It also included attempting to ensure stability in the workplace to keep the economy humming, as well as to make sure even the unemployed and even unemployable had the wherewithal to cover their consumption needs. Directly after the war, for example, there was much labor unrest, particularly in cities, as returning veterans considered wage rates to be below what was necessary for them to get ahead. During the war, factory workers consisted mostly of women and minorities disallowed from direct participation on the military front. Such workers were new and more vulnerable to pressures to keep wages low. They were also working within the general context of the personal sacrifices necessitated by the war effort. In fact, because demand remained high and wages low, profit rates in all the major sectors of the economy were extremely high during the war and the larger corporations became ever larger as a result.

To help regulate real and potential labor unrest in the post-war period, the federal government therefore jumped into the private labor market with both feet. In 1947, a major government initiative officially titled the Labor-Management Relations Act (more commonly referred to as the Taft–Hartley Act) was passed into law. This law began the process whereby labor unions in the major sectors of the economy, particularly automobiles and directly automobile related, steel, rubber, and petroleum, were given specific rights in exchange for keeping the peace among their members. Just like Ford in earlier days exchanging higher wages for labor stability, the federal government was now actively managing the process. While biased toward the demands of big business with the most powerful lobbies in Washington, D.C., the act better established the process of formal collective bargaining in good faith between labor unions and management. In this process, workers in major industrial sectors of the economy were able to establish pension plans, health and welfare benefits, some participation in the setting of work rules, such as annual and sick leave, as well as compensation for overtime work, and even consultation over the institution of new technologies that may affect their jobs. Perhaps of most importance, however, is that collective bargaining between labor and management eventually led to longer-term, mutually agreed upon contracts with regular cost-of-living increases for unionized workers. This helped to ensure the stability of production over time as well as more middle-class wages for those workers covered by such contracts.

Just like Ford's higher wages in the early decades of the twentieth century, then, rising incomes among big unionized workers in the post-war period ensured an ever-growing mass market for all sorts of consumer goods, from food to furnishings, for the new middle-class houses of the suburbs. So it was the combination of Keynesian aggregate demand enhancement via continued high government spending as well as the new role of government in managing the labor market that have led many observers to suggest that the post-war economic boom was a period of Fordist compromise among big government, big business, and big labor. Before moving on, however, it is important to note that not all sectors of the economy were union organized and therefore part of this compromise. The textile industry, still a major employer, avoided the unionization movement almost entirely, for example, as did most small- and medium-sized manufacturers and other businesses, particularly retail establishments. Also, many states, especially in the south and the west of the country, were quick to pass anti-union right-to-work laws to avoid the process altogether, a fact which will become greatly significant as the U.S. economy begins to change in the late 1960s and 1970s.

CITIES AND THE POST-WAR ECONOMIC BOOM

The political regulation of the labor market thus ensured steady growth in the economy and collectively bargained wage rates and increases produced a mass market for the mass production of the nation's factories. As other industrialized countries in Europe as well as Japan were still recovering from the social and material devastation of the war, the U.S. economy was growing steadily, largely on the basis of this expanding internal market. Indeed, this domestic growth was greatly enhanced, in turn, by the large export markets found in these still war-torn countries. As a result, major industrial cities such as Pittsburgh, Cleveland, Cincinnati, Detroit, and Chicago grew rapidly in population and continued to sprawl out into ever-new suburbs along expanding networks of roads and highways.

Big industrial corporations continued to grow as a result of economies of scale and mass production techniques. Mass production, in turn, created its own markets through falling prices, big government ensured even more consumer demand via collective bargaining and various forms of social welfare, and big unionized labor ensured relative peace in the workplace, thereby guaranteeing that this whole process would continue without interruption. The new middle-class wages of unionized labor brought mass consumption, from automobiles to suburban homes and all the necessary furnishings. Another surge in automobile ownership took place: from approximately 26 million owned in 1945 to as many as 52 million in 1955 and 97 million in 1972, with all that this meant for demand for industrial inputs, petroleum and petroleum products, and, of course, paved roads and highways. More massive middle-class consumption also helped to drive the development of air travel and it is during this period that airport development becomes a factor in the further stretch of the built and social environments of cities toward former rural areas.

This Fordist compromise among these big institutions combined with the fact that Europe and Japan were still rebuilding after the devastation of the war made for a situation in which U.S. corporations were most focused on the domestic middle-class market. Long-term, collectively bargained contracts ensured both labor peace and a growing market for whatever was produced. Growing markets, in turn, generally meant continually rising profit rates. Furthermore, as major sectors of the economy became increasingly dominated by only a few large corporations, agreements among them over wages and prices led to far less competitive pressure all around. Finally, the whole system was based on cheap imported petroleum, which fueled everything from automobiles and trucks, to industrial machinery, to machine tools, to electricity generation.

So the Fordist compromise entailed an effectively managed post-World War II expansion of the U.S. economy, based largely on domestic mass production and consumption. As mostly suburban homeownership expanded from about 45 percent of the population in 1945 to as much as 65 percent in 1973, great industrial cities such as Pittsburgh, Cincinnati, Cleveland, and Detroit continued to sprawl widely over formerly rural territory. Similarly, the continual rise in automobile ownership ensured that road and highway building continued at a rapid pace, leading to much enhanced mobility for city-dwellers and business alike. Thus, mass production/mass consumption and suburban sprawl fed off each other in an increasingly rapid self-propelling manner.

ECONOMIC CRISIS AND THE TRANSITION TO POST-FORDISM

By the mid-1960s, however, difficulties began to emerge that eventually broke the Fordist compromise in the 1970s. First, Europe, particularly Germany, and Japan had virtually recovered from war devastation and were beginning to compete with their industrial products in the large domestic market of the United States. By the mid-1960s, industrial products from Europe and Japan had attained levels of quality similar to those produced in the United States but were much less expensive, largely due to lower labor costs. This put these new imported products directly in competition with domestic ones for the first time and even some of the newest industrializing countries like Taiwan, Hong Kong, and South Korea were beginning to compete in the U.S. market with lower-priced commodities.

Second, this new competition for domestic markets was exacerbated greatly by the oil crisis of the early 1970s when oil prices increased dramatically and quite abruptly. The newly formed Organization of Petroleum Exporting Countries (OPEC), dominated by major producers such as Saudi Arabia, Iran, and Iraq, raised prices unilaterally because they could not get an agreement from major oil-consuming states for better prices for their natural resource. This rapid increase in the price of oil had a devastating effect on the economies of major oil-consuming countries. Automobiles and trucks produced in the United States, for example, were constructed on the assumption that oil prices would remain forever low. They were built heavy and big with much more attention to style and size than overall fuel efficiency. As oil prices rose dramatically, the infiltration of European- and Japanese-built vehicles into the U.S. market was greatly facilitated as such vehicles were smaller, lighter, and much more fuel efficient. Thus, the new foreign competition came to be felt most keenly in the important, still labor-intensive automobile industry.

But, significantly, petroleum was also used for power in all sorts of other machinery, as well as in the overall generation of electricity. Because of this, rapidly rising oil prices were not just felt in the automobile industry and its various sub-sectors, but throughout the economy as a general energy cost crisis. The cost of production of commodities rose dramatically across the economies of industrialized countries, requiring much attention to cutting other costs of production to offset this rise in energy prices. And herein is the source of an important wedge which will eventually break the post-war Fordist compromise completely apart: the cost of labor, another major cost of industrial production across the board, had now to be taken into very careful consideration for potential savings.

Briefly, this newly necessary and thoroughly concerted effort to cut labor costs led to a threefold outcome important to cities. First, workers were asked to open up new negotiations of already collectively bargained longer-term contracts. This led to new and sometimes severe labor strife in cities in the form of strikes, lockouts, and threatened abandonment of factories by owners, particularly since workers were given little choice but to negotiate. Second, major corporations put new emphasis on labor-saving technology in the production process which eventually resulted in innovations in labor-saving, increasingly computer-aided, automated production processes. Third, corporations now began in more earnest to look for locations of lower-cost labor, both within the United States, particularly in the less urban and industrial southern and western **Sunbelt**, and increasingly to other even lower-wage countries to move their factories.

In short, growing international competition combined with the energy crisis broke apart the post-war Fordist compromise. Increasingly, long-term collectively bargained contracts for labor as well as big government welfare state and regulations of the workplace and the market were considered a drag on the ability of business to react to such fast arising competitive challenges. Rather than poor business decisions and the lack of longer-term planning for a changing marketplace on the part of large private corporations, in other words, it increasingly became common propaganda and then common wisdom that competitive failures in the U.S. economy were the fault of overly bureaucratic labor intransigence and too extensive and overly rigid government regulation. These were considered the main sources of the lack of flexibility on the part of major corporations to meet such new challenges in the marketplace.

TOWARD POST-FORDIST, POST-INDUSTRIAL CITIES

The key term here is "flexibility." Corporations, it was argued, needed to be able to meet competitive and energy cost pressures as rapidly as possible. Big labor and big government needed to be more flexible with their already negotiated contracts and regulations to allow this to happen. This new common sense entailed, then, a newfound faith in private corporate decisions and free markets as opposed to social compromise and government regulation as a means to ensure continuing economic success. Thus began what has been called the post-Fordist period when corporations began to make more rapid and unfettered decisions as to how to produce and, importantly for cities, where to produce their products. New labor-saving technology was introduced where possible and new locations of production were more actively sought, as already mentioned. This led to an overall decline in the number of city-based manufacturing jobs, as well as a transition from more skilled to less skilled for those jobs that remained as a result of the spread of automation.

For industrial cities such as Detroit and Pittsburgh, this meant that jobs were lost, factories abandoned, and overall economic prospects became increasingly bleak. The loss of manufacturing jobs meant a self-propelling downward spiral multiplying throughout the entire city economy as workers could no longer consume to the extent that they previously had and, as a result, many more lost their jobs in wholesale, retail, and other businesses. This decline of industrial city centers was only made worse by automobility and the ongoing construction of roads and highways. Those businesses and people who could sought new locations in the suburbs and the suburbs of suburbs, or **exurbs**, in a seemingly endless process of sprawl. What was left in the inner cities were increasingly the least skilled, the least educated, and the most poverty-stricken people living in increasingly derelict housing surrounded by increasingly abandoned factories and warehouses. And in the context of the United States, this meant that an increasing proportion of people still living in the inner city were minorities, particularly African-Americans.

The post-Fordist quest for corporate flexibility also entailed major changes in the organization of production and the distribution of products. In the Fordist period, the few major corporations which came to dominate each major sector of the economy oversaw and controlled each stage of production, from raw material to final product. This made for gigantic corporate organizational structures that were rigidly bureaucratic and resistant to change. New competitive and cost pressures necessitated less rigid ways of doing things. Corporations thus began to subcontract some of the stages of production to other firms, including financial services, research and development, and marketing. As described, the CBDs of major cities were already in the process of changing from majority manufacturing to majority office jobs as major corporations increasingly located their headquarters there. Now these new financial and other professional services firms also came to locate in CBDs, along with major banking institutions. This widened the skills mismatch between those who actually lived in the inner cities and those overwhelmingly suburban professionals who commuted daily to their office jobs in the CBD.

All told, the quest for flexibility in production and distribution, along with innovations in transportation, now including air travel, and communications, now including FAX machines, email, and the Internet, created a situation whereby large corporations could coordinate the entire production process across a number of different locations. The new, flexible post-Fordist corporation can locate its headquarters in major centers such as New York and Chicago, its research and development facilities in major centers of innovation such as Silicon Valley or the Research Triangle of North Carolina, its low-skilled customer relations center in North Dakota, and its equally low-skilled, automated assembly plant in low-wage Alabama or even lower-wage Mexico. With the modern interstate highway system and a fully developed network of airline travel and airports, any raw materials that are needed can be delivered and markets for the final product can be supplied just about anywhere. Indeed, add to this scenario innovations in multi-modal containerized shipping, trucking, and rail, and it is easy to see how this process quickly became global as more and more U.S. corporations found even lower wages even farther offshore than Mexico.

This post-Fordist flexibility of production and location had enormous impacts upon cities, as will be described further in the following chapters of this book. At this point, what needs emphasizing is that the dual crisis of increasing international competition and rising energy costs of the 1970s made for a situation in which not only city economies but also the domestic economies of all industrialized economies were beginning to transition from majority manufacturing to majority service-sector jobs. This is what is meant by those who now speak of **post-industrial cities**. City economies are overwhelmingly dominated now by professional office jobs and other service sector jobs in retail and entertainment. And, importantly, this is also the case in lower-wage cities and regions of the Sunbelt, which initially attracted manufacturing jobs away from northern industrial cities as a result of the continuing automation of industrial assembly overall.

THE PLIGHT OF POST-COLONIAL CITIES

The following chapters will explore this notion of post-industrial city life in more detail. Before going there, it is important to note that all former imperialized countries of the Global South attained formal independence during the post-war years, although some, such as Angola and Mozambique, not until as recently as the 1970s. Independence usually came at the cost of open conflict and the destruction of much important economic infrastructure, particularly in major cities. Such conflict also brought forth much antipathy not just between indigenous groups and European imperialists, but also among indigenous groups themselves, forced together, not by choice, within national boundaries created by Europeans. Indeed, this antipathy (and even outright conflict) lingers today, particularly in more recently independent countries in Africa where the legitimacy of indigenous rulers is open to serious question by other groups within such boundaries bequeathed by imperial force, such as in present-day Sudan.

Even more serious is that these newly independent post-colonial countries have been economically molded by the specific interests of European imperial powers. From predominantly cash-crop export orientation to the creation of funneling transportation infrastructure, on to the overwhelmingly primate city structure, independence was, in most cases, only formal and not real. Most countries are still trapped by longstanding global production and trade patterns as raw material and semi-processed goods producers for final production in advanced industrialized countries in the

Global North. Attempting to break this dependent economic status would take concerted effort, to be sure, and many argue that most former imperialized countries have not been very successful at this, remaining forever behind in the quest for truly independent economic development status. While former imperialized countries have achieved formal independence, in other words, many remain essentially in a neocolonial relationship with former imperial countries.

Indeed, post-colonial city structure and function in the Global South continues to exhibit most of the traits of the former imperial relationship. Primate cities have continued to grow rapidly both because of general population growth as well as the lack of adequate opportunities in the countryside. Attempts to reorient economies toward a more internal focus have stalled in most cases because of the lack of both resources and political will. In the end, as will be discussed in more detail in the chapters to come, post-colonial cities have come to exhibit the following characteristics:

- primate city enormity in which the largest cities actually have increased dominance over the next-sized cities, leading many now to talk of **megacity** development;
- rapidly growing problems, both environmental (air and water pollution, waste disposal, etc.) and social (growing polarization of wealth and poverty, shanty town and ghetto formation, lack of security, etc.), that come with such over-urbanization;
- a continuing coastal port orientation of such primate megacities even after concerted efforts on the part of some countries, like (most famously) Brazil, to relocate economic and political functions toward the interior of the country;
- continually rapid population growth and rural-to-urban migration which merely exacerbates the problem of primacy but has also led to growing diversity of city populations; and
- at the other end of the former imperial relationship, a continuing and sometimes rapid increase in the numbers of immigrants in imperial countries from their former imperial possessions.

These new immigrants were both solicited and came on their own volition to fill the labor needs of receiving countries, particularly for semi- and unskilled jobs. This greatly increased the ethnic diversity of particularly European cities leading to all sorts of changes in their cultures, both in terms of peaceable fusion and of more conflict-ridden standoffs over such things as equal opportunity and outright discrimination and ghettoization.

Box 4 Cities of Sultans and Mughals in India

Rural pursuits and small village life was the main mode of existence on the vast Ganges plain of what is now India for many a long century. Regional kingdoms grew and declined during these centuries, but their capital cities were neither large nor, in many cases, very long lasting. This situation began to change during the thirteenth century CE, however, with the incursion into the subcontinent of Turkic Muslims from Central Asia. Their foray southward into the subcontinent resulted in towns eventually being rebuilt as fortress-cities in Lahore (now in Pakistan), Delhi, Agra, and elsewhere to house the Turkic leaders of the conquest. The planning and architecture of these cities reflected the need for protection against the periodic unrest on the part of the conquered people, by far the majority of the local population. Massively monumental and quite central and dominant fortresses dominated the skyline and housed the elite within city walls. These cities were rebuilt over earlier city structures as to the dictates of Islam, including strategically located and thoroughly monumental mosques for community worship, as well as large shrines for departed holy people and other revered ancestors. Indeed, Islam was, and remains, quite conducive to city life, in general, both because of its specific teachings which led to the original conversion of city-based Arab merchants as well as given the religion's requirements of regular weekly worship in the company of other believers and community reverence for holy shrines.

Yet, it was not until a further wave of Islamic peoples from North and Central Asia migrated in greater mass southward into the subcontinent during the fourteenth century that town and city life began to grow and spread more regularly in this part of South Asia. These Central Asian peoples created the Mughal Empire (1526 to 1857) that eventually covered most of the Indian subcontinent.

Mughal authorities, like earlier sultans, took a particularly keen interest in city building and architecture. City fortresses were expanded and embellished, new monumental central and neighborhood mosques were constructed, gardens and waterworks were built, and gigantic holy shrines proliferated to dominate Mughal cityscapes. Perhaps easiest to imagine as an example of Mughal city monumentalism is the white marbled Taj Mahal, which was built by Mughal Emperor Shah Jahan as a memorial to his deceased wife, Mumtaz Mahal, between 1632 and 1653 in Agra.

It was Shah Jahan who established Delhi (Shahjahanabad) as the dominant capital of the greater Mughal Empire, which led to major rebuilding and expansion of the city's fortress and other central areas in this monumental style. He also undertook similar major rebuilding in the regional capitals of Lahore and Agra. Indeed, soon after Shah Jahan established his court in Delhi in 1648, the population of the city is thought to have been as high as 400,000 or so.

In short, the Delhi of the Mughal Empire reflected the dominant role it played as the residence of the emperor who ruled over a vast agrarian empire that was populated overwhelmingly by people who were both ethnically different and followers of different religions. The cityscape of the capital and, indeed, of all the major cities of the empire had to be monumental on a gigantic scale to reflect the power of this conquering minority, as well as that of Islam. And so Mughal city monuments continue to be predominant features of many an Indian city to this day.

FURTHER READING

A good general overview of suburbanization in the United States is provided by Robert Beauregard in his *When America Became Suburban* (Minneapolis, MN: University of Minnesota Press, 2006). A broad account of the first Industrial Revolution in Britain from the particular point of view of gender relations is found in Joyce Burnette's *Gender Work and Wages in Industrial Revolution Britain* (New York, NY: Cambridge University Press, 2008). Friedrich Engels provides a classic look at *The Condition of the Working Classes in England* (Stanford, CT: Stanford University Press, 1968, original 1845) in the midst of rapidly growing industrial cities. A very good account of the experience of imperialized cities of the Global South can be found in Anthony D. King's *Colonial Urban Development: Culture, Social Power, and Environment* (Boston, MA: Routledge and Kegan Paul, 1976) and Steven King provides a recent overview of industrialization in his *Making Sense of the Industrial Revolution* (New York, NY: Manchester University Press, 2001). Finally, Steven P. Blake provides a detailed account of the reemergence of the Indian city of Delhi (Shahjahanabad) as the capital city of the Mughal Empire in his *Shahjahanabad: The Soverign City of Mughal India, 1639–1739* (Cambridge: Cambridge University Press, 1991).

5

CITY ECONOMICS

The preceding historical chapters have made clear that, after the Western Renaissance and the spread of post-Renaissance culture across the globe as a result of Western imperialism, city location, morphology, and overall development have been based fundamentally on economic criteria. As new trade routes opened up, as new technologies and organizations of production arose, and as innovations in transportation modes and networks came to be, the fortunes of individual cities changed, sometimes quite dramatically. This remains even more the case today, in this age of globalization where Western merchant culture has deepened its hold on most every part of the planet. Even formerly Ancient/Classical cities, through the age of imperialism on to today, have been reorganized according to economic criteria so that social and built layers of earlier city culture, while still noticeable, are more or less smothered by office towers of CBDs and industrial and post-industrial urbanized bits scattered in ever-expanding metropolitan areas. Indeed, traditional, symbolically important cities such as Beijing are literally losing their status in urban hierarchies to new, very rapidly growing economic centers in cities such as Shanghai, as well as Guangzhou and others rapidly sprouting along the southern coasts of China. Economic development is the essential key to evolving city life around the globe as policy authorities in more or less every city aspire to achieve **global city** status for their particular site. And, as a result, all cities, both in the Global North and the Global South, are beginning to look alike, with the same sorts of built and social environments and, indeed, the same sort of city amenities, from luxurious convention centers to festival marketplaces for high-income consumers to state-of-the-art sports stadia and high-rise office structures, all built in similar fashion with similar materials.

This chapter addresses the reason behind this seeming globalization of city development by focusing more closely and specifically on city economic processes. As to this, the most important issue for post-Renaissance cities is the actual location of economic activities and, then, people, both as producers and consumers of goods and services traded in the commercial marketplace and as residents of particular areas of the city. During the Mercantile/Colonial period, as we have seen, the economic success of cities was very much contingent upon their location as easily accessible ports, along inland waterways and on the coast. The key term here is accessible. These were locations in which goods from afar could be easily landed and traded in close proximity to the port. Again, the transportation of goods (and people) was far cheaper by water at this time than overland. Thus, in turn, the extent of the market for trading goods in the interior of countries depended upon good inland waterways and, in the case of Europe, the increasing ability of trading ships to sail out of the Mediterranean Sea toward northern and eventually Atlantic Ocean ports elsewhere in Europe.

But note that accessibility is not just a factor for city development for this time period alone. Once economic criteria become the most important element in the growth and decline of cities, focus needs to be put continually on changing transportation modes and networks. New shipping technology as well as increasing knowledge of shipping routes and, indeed, the Earth's territorial extent, for example, led directly to the relative economic decline of the original merchant cities of Italy and the rise of Atlantic Ocean ports such as Seville, Lisbon, and eventually Amsterdam and London. Similarly, when the port of Bruges in Flanders became less and less accessible to oceangoing ships because of a growing problem of silt, Antwerp and Ghent became more economically viable. Finally, note that water transportation can also be facilitated by human constructions.

Canal building, for example, provided a means to artificially extend the market for goods inland. Canals also shorten shipping routes more globally, as most famously in the case of the Suez and Panama canals, which led, respectively, to the reinvigoration of the economies of Mediterranean ports in Europe and an early, quite significant, means to bind the economy of the United States from the earlier established Eastern seaboard port cities to newer cities arising on the Pacific Coast.

As discussed in Chapter 4, this locational logic can be followed on into the Industrial/Imperial and post-Industrial/post-Colonial periods of city development as modes and networks of overland transportation became more efficient first along fixed rails and then almost anywhere along roads and highways and, eventually, on the basis of air travel. What needs underscoring here is that each such innovation in transportation leads to the rise or decline of the economic viability of particular cities. It also leads to changes in the social and built morphology of cities manifested most spectacularly in the continuing stretch or sprawl of metropolitan areas outward from original city centers.

WHY DO BUSINESS FIRMS LOCATE WHERE THEY DO?

So much for the economic viability of the actual location of cities; now it is time to think more carefully about the location of economic activity itself. Mercantile/Colonial cities were obviously dominated by merchants, either distributing the goods procured from afar or managing the actual procurement of such goods from growing colonial possessions. The most important, most profitable merchant houses tended to locate their offices and warehouses as close to the actual port as possible. Again, accessibility was the key, in this case, for more efficient offloading cargo and reloading for sale in the city or in accessible markets outside the city. The prime port location of such firms thus reinforced their profitability, allowing such firms to remain dominant in the local marketplace.

So, actual location within the city can provide an important economic premium to firms. Notice now three further implications of this. First, such "prime" locations will be much sought after by firms for this very reason. Firms essentially will compete for such locations, which thereby bids up the rent or sale price for such sites so that each firm must calculate whether or not the economic premium it gets from such a location offsets the actual cost of such a location. Essentially, this economic premium itself is not fixed for all time but changes with variations in, for example, what is traded and how, how and by what means goods are produced, and, of course, changes in transportation modes and networks. A location that once offered an economic premium may in the future actually produce an added cost of production due to such changes. The high land rents of port locations, for example, eventually proved too costly for merchant firms facing competition from firms increasingly able to locate in lower-rent districts away from the port, yet still able to efficiently offload and upload their cargoes because of innovations in transportation modes and networks. The same thing happened to automobile manufacturers in central city locations facing growing competition from firms located in much cheaper suburban sites after World War II. And, again, the land-use pattern within cities generated by this bid-rent process for prime city property renders the familiar visual pattern of city landscapes growing taller in an attempt to spread the costs of such a prime location the closer one gets to the city center. Indeed, as cities have sprawled into the present due to automobility, one can now discern prime locations variously dotting entire metropolitan areas on the basis of the verticality of the built environment, even in suburban and exurban areas.

This last point also signals the third implication of this notion of prime location within cities. Since primacy comes with higher, and most times much higher, rents there is a strong incentive for businesses to find a way to locate elsewhere and still compete with those which have outbid them for such prime locations. This was a major although, of course, not the only factor in the spread of Renaissance merchant culture to more and more cities. It also continues to be a major factor contributing to urban sprawl. The higher costs of doing business in prime locations become, in this respect, a potential drag on the competitive status of those businesses located there. If it is possible to do the same business from less costly, less prime locations in the suburbs and now exurbs, then the competitive edge will go to those businesses which take advantage of this opportunity. And notice how this becomes a self-propelling process: as suburban and exurban businesses become more and more possible with road and highway development and increasing automobility within the metropolitan area, they will eventually out-compete those businesses still located in high-rent, formerly prime locations toward the center of the city. As a result, two things happen at once. First, certain kinds of business activity such as mass lower-end wholesale and retail and manufacturing will find it increasingly uncompetitive to pay such high central city rent, while others such as high-end retail, professional office work, and banks and other financial institutions will take the place of such businesses in the CBD. Second, of course, metropolitan sprawl will continue to spread as new prime locations in the suburbs emerge and then eventually become high rent, and new prime locations emerge as suburbs to suburbs to suburbs, etc.

WHY DO SIMILAR FIRMS CLUSTER AROUND EACH OTHER?

Another aspect of business location important for city development is the tendency for like businesses to locate in close proximity to each other. For example, it is just by chance that Henry Ford started his automobile business in the early years of the twentieth century in Dearborn, Michigan, just outside of Detroit. As his business thrived, however, Ford established and maintained close connections with suppliers of needed raw materials for the manufacture of automobiles and eventually established retail outlets for the finished product. At the same time, Ford's new organization and technology of mass production in his manufacturing plants were molding a workforce into mass autoworkers. Combining all of this with Ford's social policy of higher wages to ensure industrial peace, as well as mass markets for his product, the Ford Automobile Company established a full system of production and consumption from iron and steel to assembly-line final production onto the retail sales floor of automobile dealers.

Indeed, up until the economic crisis of the 1970s, most large corporations sought direct control of all the phases of such industrial production from, for example, Ford-owned iron and steel plants to Ford-owned glass producers to Ford-owned assembly plants to Ford-owned retail outlets, with each stage connected by Ford-owned transportation modes. This **vertical integration** of companies such as Ford was to ensure close control over the quantity and, most significantly, the quality of inputs and outputs on through to the final product for sale. During the Fordist compromise of the post-World War II era, most every sector of the economy became dominated by a few large companies precisely because of this ability to control directly all phases of production. As discussed in Chapter 4, however, such corporate gigantism became one of the main impediments to necessary change within the new competitive environment of the 1970s.

Of importance here is that as the Ford Motor Company thrived in and around Detroit, it literally created a metropolitan environment of industrial supplies network, labor force, and retail outlets most conducive to the production and consumption of mass-produced automobiles and trucks. It also provided much impetus to the development of roads and the growth of automobile-related businesses such as repair shops, specialized machine tool providers, and gasoline retailers. In the end, the existence of such an automobile-oriented metropolitan industrial environment in Detroit made it very attractive for others hoping to compete in the automobile industry. The key here is that the second automobile company which comes to locate in the Detroit area will actually enjoy what are called agglomeration economies precisely by making the decision to locate there. This second company will not have to start from scratch, so to speak, in creating its own automobile-related supplies network, labor force, or retail patterns. It can compete for the services of already formed relationships. The third automobile company to locate in Detroit will also enjoy such agglomeration economies as the automobile-related industrial environment of the metropolitan area has been actually enlarged by the activities of the second locator. And so Detroit eventually became "Motown" (Motor City), as first General Motors came to locate there and then Chrysler was established.

Automobile companies thus came to cluster around each other in Detroit in order to take advantage of agglomeration economies. Such economies accrue to all second and further locators in any industry for the same reason. As a result, cities become dominated by specific industries as similar companies clustered around each other for competitive reasons. But, importantly, another aspect of agglomeration economies concerns keeping up with the competition. Spatially clustering around directly competitive firms allows for the possibility of keeping one's eye on what the others are doing in order not to be left behind by the latest innovation in the organization ortechnology of production. This further benefit of spatial clustering not only helps to explain why whole cities come to be dominated by specific industries, but also, importantly, why specific districts within cities come to be so singularly dominated, from garment districts to financial districts to more contemporary high-tech districts.

But agglomeration economies can eventually turn into what are called **diseconomies of agglomeration** for similar reasons associated with bid-rents. Many companies within a particular industry may want to locate in close proximity in the finite territory of the city or in districts within the city. This will, as we have seen, drive land rents and, then, overall production costs up. Furthermore, increasing numbers of businesses clustering within the city will mean that much more traffic on city streets and that much more densely packed working-class housing and people occupying city territory. Again, these increasing diseconomies of agglomeration help to drive urban sprawl as businesses eventually seek more competitive locations outside central cities, yet still accessible to city markets.

THE CHANGING LOCATION OF CITY PEOPLE

Another major aspect of city economies concerns where people come to be located. As described in previous chapters, through the Mercantile/Colonial era, the richest merchants both worked and lived in the prime central locations of the city. This changed in the Industrial/Imperial era with suburbanization and urban sprawl; but, interestingly, this sort of central city location of the rich is the case again in the post-Industrial/post-Colonial cities of today. Before describing this latter it is important to note again that the industrial era really was the turning point in city development as it created a continual process of rural to urban migration. This was the time when large multistoried industrial factories sprouted up in the middle of cities employing growing numbers of cheap low-skilled workers. Housing this new, rapidly growing industrial working class in cities thus became a huge issue. The inner-city residential areas surrounding factories in the CBD came to be generally and sometimes highly overcrowded. Formerly single-family houses of rich merchants and other earlier city elite were increasingly broken up into multifamily dwellings. Other multifamily dwellings were hastily and most often haphazardly built to meet the growing needs of the new working class. Each individual worker had very little money from meager wages to pay rent. The total rent money of many workers crowded together, however, could be a source of good profit for city property and building owners. Thus, there was a large incentive to build upward as cheaply as possible and to crowd as many into residential buildings as they could hold. In this respect, low wages, overcrowding, and slipshod housing became inherently linked phenomena in the inner city.

So the inner city became increasingly the location of working-class housing. Here it is important to note that increasing numbers of city people, even of working-class status, create a market for other things besides housing. Everything from groceries to clothing to furniture to household appliances and so on were now needed by the new city-based working class. This created many more jobs for people providing these commodities who then spent their money on similar things and so on. At the same time, janitors, plumbers, and other tradespeople, as well as those in the professional services, from banking to legal professions, found growing city-based markets for their services. Finally, restaurants and other entertainment establishments also found growing markets leading to more jobs, and on and on. In this way, the economic development of cities becomes a self-propelling process.

Yet, this self-propelling engine of growth can only be maintained if the major employers of any particular city continue to hire more and more people. Once such an employer falters or, indeed, moves its major operations to the suburbs or, today, increasingly to other countries, this upward spiral of growth can quickly turn to a downward spiral of economic decline. In that case, a multiplier of job decline can be traced in the same manner as the multiplier of job growth was traced in the previous paragraph.

SUBURBAN DREAMS OF THE MIDDLE-CLASS GENTRY

Indeed, such a vicious cycle of decline came to occur in most every industrial city in the post-World War II era. As streetcar suburbs expanded into highway infill suburbs, more and more of the middling classes sought the suburban dream. The residential population of inner cities thus became increasingly dominated by the lowest classes unable, even with the help of government-backed mortgage support, to move out. For a time, the new middle- and upper-class suburbs were, indeed, bedroom communities in the sense that the major breadwinner of the household (usually male) commuted to their job still located more centrally in the city. So, early on in this suburbanization process, the jobs remained in cities.

But notice that this suburban dream of the middling to upper classes entailed three significant results. First, the male breadwinner of the household spent most of his day in more central city locations. This created a daytime market for upper-end retail and entertainment goods and services, particularly around central city commuter rail stations and in the CBD. Second, as commuting shifted from rail to automobile, central city streets became extraordinarily congested and adequate parking almost non-existent. This traffic congestion and parking problem made doing business in the city much more difficult. In turn, the parking problem made it increasingly difficult for big department stores to cater to middle- and upper-class markets in their multistoried center city locations.

The third result of the suburban dream was that new single-family houses in formerly rural areas had to be furnished, to be sure; but they also were occupied all day by those non-commuting members of the household, particularly, of course, women and children. Food, clothing, and other consumer goods were necessary to supply them with day-to-day needs. It did not take long for retailers to figure this out as more and more businesses sought suburban locations to meet such needs, leaving behind formerly central city sites. The most spectacular manifestation of this trend is the suburban retail mall complex so ubiquitous in the American landscape. Indeed, the malling of suburbia solved two problems at once: it brought needed retail outlets to rapidly growing suburban populations and it solved the parking problem, offering generous and free parking on acres of paved flat top surrounding such outlets.

Increasingly, then, it was not only industrial firms which were leaving central city locations but all sorts of businesses, particularly those catering to the middle and upper classes. This added even more job opportunities in suburban locations, thereby accelerating the self-propelling job and prosperity multiplier effects, now quite outside city centers. Soon professional services such as finance and accountancy, legal, secretarial, and certainly real estate firms sought suburban office locations. Suburban CBDs therefore came into existence with their own "suburban" bedroom communities, eventually necessitating their own retail establishments and so on, and so on, resulting in the now ever-expanding exurban sprawl of major cities around the world.

CITY DEVELOPERS, HOUSING GATE KEEPERS, AND THE DECLINE OF THE INNER CITY

This self-propelling suburban sprawl of people, jobs, and business left little opportunity for those left in the inner city to meet their day-to-day needs, both in terms of jobs and access to adequate retail outlets. As suburbs and exurbs continued to enjoy upward spirals of self-propelling growth, inner cities generally experienced downward spirals of vicious economic decline. Abandoned factories and other physical plant, vacated department stores, and housing left to decay for lack of profitability by now mostly absentee owners, came to dot inner cities of most every former industrial city. The more abandonment took place, the more what was left was devalued by being surrounded by an increasingly derelict built and social environment. Jobs were lost, money was not available and, because of this, many more jobs disappeared, so much so that one American scholar suggests that most inner cities in the United States are now populated by an **underclass** of people, mostly unemployed and living off the dwindling largess of the welfare state.

While this last notion remains controversial, it is true that as middle- and upper-income residents left for suburbia, it severely leveled the class structure of inner cities to the lowest, less-skilled, less-educated people. But it is not just about personal or household means. It is also about the supply of adequate housing and here we return to the first part of the story about industrial working-class housing. As discussed, even when times were good, working-class housing was most hastily put up and on the cheap. This was based on conscious decisions on the part of city property owners and building developers to squeeze as much profit out of such housing as possible. The key here is that such city developers seek profits from their efforts at creating residences. Just like Ford, they both cater to, and help create, the market for housing. The decisions they make thus determine both the supply of housing and its location. As the industrialization process continued in cities, developers began to cater to the growing middle classes toward the outer territorial extent of the inner city with more spacious, less densely packed apartments and flats and even small single-family houses. Farther out toward the suburbs they developed even larger residential spaces and single-family homes and so on.

While governments in Europe and elsewhere more closely regulate housing development, in the United States it is mostly left to the private sector. Because of this, it is easier to see how private markets for housing work. Typically, the pattern for occupancy in the United States has been one in which people moved out the spatial scale as they moved up the income scale. This is not just an issue of wealth, but also that of the generally adverse perception of cities and city life among most Americans. In any case, the key to the demand and supply of housing in this process was that, with the growing general prosperity of the post-war years, there developed **vacancy chains** of housing as households progressively moved up the income scale. Starting with the smallest cheapest housing in the inner-city households, city dwellers, as their family income grew, would eventually move to larger houses farther out, to even larger houses even farther out in the suburbs, to even larger houses farther out, and so on. Each time a family left its apartment or house for a larger house farther out toward the suburbs, it opened up this former residence for the next generation of household movers with rising income to occupy.

But this vacancy chain process has never worked quite as well as the theory suggests, even in the privatized housing market of the United States. First, there has always been discrimination in housing markets, sometimes along class lines and more often along ethnic lines, and most often along a combination of these two lines. Here is where the so-called **gatekeepers of housing** supply come into play. Financial institutions and their agents play a huge role in both what housing is built and who gets to occupy it. High down payments for mortgages, for example, used to preclude most people of middle-class status and below from owning adequate housing. Financial agents can also determine where funds are made available in the city. In the past, for example, property owners in some parts of cities were overtly denied funding for the upkeep of their property and potential buyers in these areas denied mortgages outright. In fact, such discriminatory practices were so overt that financial firms could display big maps of the city with entire districts circled with red marker as absolute no-loan zones: hence the term **redlining** was coined in reference to such practices. With no funds available for upkeep and possible sale, property owners had no incentive to maintain their properties in such redlined areas of the city.

Second, from the very beginning of streetcar suburbanization on through the onslaught of mass suburbanization, other gatekeepers such as real estate agents, property managers, and even developers themselves have practiced outright discrimination via ethnic restrictions against, for example, Jews and people of color. Suburban and exurban homeowners' associations have also created restrictive housing codes that discriminate against larger or blended generational families by not allowing certain kinds of modifications to homes built explicitly for the stereotypical married-couple, two-child family.

Most recently, another factor has impeded what vacancy chain process actually exists for city housing. This involves the twin phenomena of the **gentrification** of major parts of inner cities and the lengthening and even breaking of the last vacancy chain link in the suburbs. Gentrification is a process by which some members of the middle- and upper-income classes (the "gentry") return to more central city locations to live and do business. In the process, they buy up and remodel residences as well as refurbish central city business areas to cater to their needs and those of their similarly situated new customers. In most cases, this gentrification of inner cities is displacing the mostly lower-income original inhabitants via higher rents, the reversion to single-family housing of formerly multifamily dwellings, much higher home values and property taxes, as well as active discrimination by housing gatekeepers. Gentrification thus tends to break the vacancy chain at its very beginning: that of the previously cheapest housing available. It also puts upward pressure on the cost of the remaining cheap, affordable housing in the central city because of rising demand from displaced lower-class residents. And such higher costs displace even more people from the central city housing market, particularly those with the lowest income.

At the other end of the residential vacancy chain the trend in the private housing market in cities worldwide has been toward building ever-bigger, more costly houses in the suburbs or exurbs. From the point of view of private developers and builders, it is much less risky to build and sell one or two huge and pricey homes than several smaller more affordable homes. The profit from the sale of one such **McMansion** is as great as or greater than the sale of, say, three or so smaller homes. In much the same way that profit is much greater and less risky for automobile firms to sell one large sports utility vehicle (SUV) as opposed to several compact cars, property developers and builders thus opt for McMansions. But the result is that the step up from the less grand, still large suburban home to a McMansion is huge in terms of the relative growth in family income. And so families who would normally forge the next link up the housing chain cannot. This, in turn, decreases the vacancy rate at their level, stopping those who would normally take over their larger suburban house and so on back down the vacancy chain. In short, such private decisions concerning the supply and demand of city housing in the United States has led to a serious lack of affordable housing in cities and, indeed, a lack of housing altogether in the direst situations.

INDUSTRIAL RESTRUCTURING AND THE DE-INDUSTRIALIZATION OF CITIES

Chapter 4 began the discussion of the breakdown of the Fordist compromise among big government, big corporations, and big organized labor. Here more details of this are needed to understand the impact upon cities of this transition to post-Fordism. The twin crises of increasing international competition and rapidly rising energy prices, for example, put all businesses on the quest for ways to meet growing competition, particularly by reorganizing production processes and cutting costs. Again, the key for firms was to become as flexible, and economically efficient, as possible to survive in the new context of global competition. Since energy costs were rising beyond the control of individual businesses, production and operations cost savings had to be found elsewhere, especially big ticket costs of production such as labor and rent.

The high rents being paid for prime central city business locations thus became an important factor in the competitive calculus of firms. Indeed, factoring in these rents rendered central city locations more and more uncompetitive. In short, firms increasingly looked to the suburbs and even to rural areas for cheaper land on which to build whole new facilities outfitted with wholly new labor-saving technology or organizational layout. The key here is that it is almost always cheaper to build wholly new facilities with wholly new technology and organization on greenfield sites than it is to try to retrofit older facilities with such.

So, again, for production cost savings, firms began in earnest to locate first in the suburbs and then in the Sunbelt and then, most recently, in even lower-cost countries of the Global South. At the same time, larger vertically integrated corporations such as the Ford Motor Company were looking for more flexibility in their organization of production. Thus began a trend toward the **vertical disintegration** of firms in which major functions such as accountancy, legal services, staffing, finance, and other managing services, as well as even parts supply, raw material procurement, wholesale/retail functions, and maintenance/janitorial services, were effectively subcontracted out to other firms. Indeed, this **out-sourcing** of previously corporate-controlled functions is a common way to do business in today's global market and is extremely important for city economies because most of the new small- and medium-sized professional and management services firms tend to cluster around each other in post-industrial CBDs, as noted in Chapter 4.

Finally, large corporations seeking cost advantages began to reorganize their whole production process according to geographic advantage. For example, labor-intensive assembly plants were increasingly located in regions and eventually other countries with large supplies of cheap, low-skilled labor. The same became the case for call-center customer relations offices. Cities that were attractive for such **back office** work saw an influx of generally low-wage jobs and, as a result, eventually became branded as back office locations. Research and development centers were located elsewhere, usually in major metropolitan areas with several institutions of higher learning and generally more educated populations. The same can be said for the location of financial services that remained in-house. Finally, the headquarters of major corporations increasingly became, and remain, clustered in a few major cities such as Tokyo, London, New York, and Chicago, rendering them what some now call truly global cities as powerful control centers for the entire global economy.

Coordinating and managing such **industrial restructuring** was made possible in turn by innovations in transportation and communication like highways, containerized shipping, airlines, and FAX machines and internet communications. In fact, such innovations are now making it possible to offshore not only manufacturing and low-skilled call-center jobs but even high-skilled professional services to lower-cost regions, particularly in the more formally educated countries of the Global South like Brazil, India, China, and South Africa. In the end, the effects on cities of these changes in corporate production and organization have been tremendous. Again, in city after city in the Global North the vast majority of workers now work in the service sector as their economies have been effectively de-industrialized.

A SPATIAL MISMATCH BETWEEN JOBS AND PEOPLE IN THE CITY

The service sector, however, is a very diverse sector. It incorporates jobs such as flipping burgers at the local fast-food outlet as well as jobs for highly educated, highly skilled people in the financial and legal professions. The key to city economics in this new post-industrial era, then, is how many of the former low-paying service-sector jobs exist as opposed to the latter high-paying ones. Indeed, this is a question for entire formerly industrialized economies because it is clear that far fewer better-paying professional jobs have been created than were lost in manufacturing in the de-industrialization process. In any case, the economies of former industrial cities such as Detroit, Pittsburgh, Manchester, Sheffield, etc., felt this de-industrialization in dramatic fashion as relatively high-paying jobs virtually disappeared and the only jobs left were in lower-end services.

For most former automobile, steel, textile, consumer electronics, and furniture workers in such cities, service-sector jobs now available either pay much less than their previous manufacturing jobs or demand more education and skills than such workers have acquired. The new lower-end service-sector jobs available also usually come with no health or retirement benefits and in most cases with less than 40 hours offered per week. At the same time, the clusters of professional services jobs such as accountancy, financial services, legal consulting and so on in the gleaming office towers of post-industrial CBDs continue to be completely out of such workers' reach due to lack of appropriate skills. This spatial mismatch of well-paying jobs in CBDs and the skills of most inner-city residents renders it that much more difficult for these latter to escape underemployment in jobs not suited for the skills they do have or, indeed, full unemployment altogether. This essentially ensures continuing economic insecurity in the inner cities; in the United States, it has also been one of the main sources of urban rioting from the 1960s to the 1990s.

Indeed, this problem of spatial mismatch and economic insecurity in post-industrial inner cities has been rendered even worse by three recent trends. First, the relocation of many retail and entertainment jobs to the suburbs or other regions has made it virtually impossible for inner-city workers to fill them because of the cost of relocation – with many already mortgaged in their increasingly devalued homes in the inner city combined with actual moving costs – and because of the lack of access to transportation, either private or public. Second, the gentrification of greater parts of the residential and business areas of central cities continues to displace already economically insecure inner-city households while replacing businesses and jobs which once catered to the needs of inner-city residents, with those now catering to the needs of growing numbers of like-skilled, like-financially enabled gentrifiers. Finally, as discussed in Chapter 6, current city development policy in most cases has focused less on rectifying the economic insecurity of traditional inner-city residents and more on creating a good business climate to attract the very sorts of high-skilled professional jobs that such residents can never hope to fill. Whether masked behind talk of progressively "creative" or "high-tech" or "global" city development, this gentrification of city development policy caters mostly to the minority of city dwellers while generally ignoring the needs of this skills-mismatched majority.

THE ECONOMIC PROSPECTS OF POST-INDUSTRIAL CITIES

With ongoing de-industrialization, the economic prospects of most cities in the Global North thus have become tied to either the attraction or generation of high-skilled professional services and the types of people who fill such jobs. What is most interesting about this is that these sorts of jobs can be located just about anywhere given innovations in transportation and communication. Even more than an industry like automobile production, such service-sector jobs mostly involve the gathering and transmission of information that can essentially flow from anywhere to anywhere else. Nevertheless, in today's city reality, one still finds such firms clustering around each other in major metropolitan areas with upper-end firms biased toward the major global cities such as London, Tokyo, and New York. The reason for this clustering behavior in the service sector has to do with still prevailing agglomeration economies of like firms, as well as more general **urbanization economies** that come with city location itself. In major cities it is easier to find firms that specialize in professional services such as accountancy, legal consultation, finance, etc., to facilitate all sorts of different business transactions. It is also easier to find a pool of higher-educated, more diversely skilled people for potential hire. Major city amenities such as better or more specialized schools and institutions of higher learning or a wider diversity of entertainment opportunities are also more likely to attract a higher skilled workforce to firms located there. These are all considered advantages for firms to continue to locate in cities: the bigger, the better. Finally, the fact that many such service-sector businesses still rely on face-to-face consultation with customers means that, regardless of the seemingly footloose possibilities for the location of such professional services, they still tend to congregate in office buildings in post-industrial CBDs.

The key for the continuing economic development of cities today, then, is, again, to attract as many such firms as possible. Such firms offer just the sort of high-paying jobs that will provide the widest possible multiplier effect throughout entire metropolitan economies. So policy-makers in most cities compete with one another for the same type of economic activity and the kind of people who fill such jobs. As discussed in more detail in Chapters 6 and 9, instead of casting a wider net or redeveloping already-existing built and social resources, post-industrial policy-makers today are largely fixated on smothering the industrial past under a façade of post-industrial amenities considered most attractive to highly skilled professionals. So, again, city after city is offering the same sort of amenities, from convention centers to festival marketplaces, for this very purpose. Indeed, this serialized reproduction of gentrified CBD sameness has led some to argue that post-industrial city economics today is all about consumption. The main economic prospect of cities is thus considered more and more to be one based on tourism and entertainment rather than the actual production of goods and services.

This seems certainly the case for those cities less successful in attracting professional jobs at the higher end of the service sector. In any case, to the extent that city development policy is based on this singular pursuit of certain types of jobs and a certain kind of labor force, those city people with different skill sets and different needs are being neglected. If all central cities are being remade in the same gentrified social and built fashion, no city can actually win the competitive development race outright as the firms and people they are trying to attract can find the same amenities most everywhere. Indeed, this situation simply allows such firms to demand even more, such as locational subsidies or outright tax holidays, from cities in order to actually locate in any specific place. Finally, aside from the fewer high-skilled professional jobs created, the majority of jobs made available in this conscious gentrification of post-industrial city economies are in much lower-skilled, lower-wage services in the leisure and entertainment and janitorial/maintenance sectors. Such jobs hardly substitute for the well-paid manufacturing jobs that were lost for the majority of inner-city dwellers in the process of the de-industrialization of city economies.

POST-COLONIAL CITIES AND THE NEW GLOBAL DIVISION OF LABOR

The economies of major cities in the Global North have thus restructured along post-industrial lines as manufacturing jobs have been increasingly offshored to lower-wage countries. Not surprisingly, this trend has had an enormous impact upon cities in the Global South. New industrial districts have arisen in cities as diverse as Kolkata, Bangalore, Mumbai, India, Jakarta, Indonesia, Mexico City, and particularly Shenzhen, Chongqing, and Guangzhou in China. Along with this new manufacturing activity has come rising incomes for a greater number of city people in poor countries, particularly among the increasing numbers of managers and other entrepreneurs taking advantage of greater opportunities in business services and the financial sector. As a result, many major cities of the Global South now exhibit an even more severe dualistic structure in their built and social environments with the new, growing, yet still relative small number of city economic elite occupying and gentrifying larger areas of central cities. Indeed, the centrally located and gleaming office towers and other gentrified urban amenities, including high-priced high-rise housing of many a city in the Global South, now closely resemble the built environment of post-industrial cities of the Global North. In turn, those who work in these offices seem to have more in common in terms of lifestyle, both material and social, with their peers in cities of the North than they do with their fellow citizens, both in the inner-city slums and much poorer suburbs and rural hinterlands. There is not much difference and much social interconnection now, in other words, between the gentrified downtown lifestyles of those in central Shanghai, New Delhi, and Rio de Janeiro and those in central Tokyo, New York, and London. In short, such built and social similarities, some argue, may actually represent the growth of a city-based, truly global culture in the making.

But the differences among cities in the Global South with those of the North are just as, or even more, important than these similarities. These small islands of gentrified similarity in the cities of poorer countries are literally surrounded by vast city spaces of extreme built and social poverty. The vast majority of city people in poor countries live in slum-like conditions with little chance of joining the gentry. Even those who are able to find work in the new manufactories are not paid wages high enough to command much city property or even much access to the new urban amenities downtown. And most other city people are even more unlucky, eking out what existence can be had in the **informal sector** of the city economy, rather than the formal sector. In short, the severity of this dualism in the built and social environments of cities in the Global South is easily apparent to anyone who ventures outside the relatively small gentrified downtown areas.

Not only this, but such dualism in cities in the Global South has been extremely exacerbated by other trends in urbanization. In the post-World War II period, a rather extraordinary phenomenon has occurred: cities in poorer countries have been growing much more rapidly than (indeed, at almost twice the rate of) cities in richer countries. Earlier in the West, cities grew in tandem with their economies. This stands to reason. With more economic growth come more jobs and therefore more opportunity in the city and so on. In the post-war Global South, however, cities began to grow quite rapidly in population; but city economies did not grow anywhere near as fast, even with the new offshoring of manufacturing from the Global North in most recent decades. Ever-greater numbers of rural people were migrating to cities in this period, then, with little or no opportunity to find formal employment.

The reason for this has to do with a wider context of demographic change in poorer countries combined with the legacy of imperialism and neocolonialism. Death rates in country after country began to fall, sometimes dramatically, after World War II, as a result of the global spread of public health knowledge and technology, including vaccines and other medicines, as well as such things as biochemical pesticides and herbicides. This **medical revolution** had an enormous impact upon the population of poorer countries as birth rates generally remained high while death rates began to fall. The rapid rate of population growth that resulted literally overwhelmed any existing opportunities for this now "surplus population" to find gainful employment, particularly in rural areas dominated, in many cases, by large-scale cash cropping and small peasant subsistence farming. At the same time, the only places of real opportunity were the one or two major primate cities that existed in most countries as a result of European imperialism. So, cities such as New Delhi, Mumbai, São Paulo, Dhaka, Karachi, and Lagos began to grow very rapidly into what are now called megacities with teaming slums of "surplus population" trying to make a living in any way possible. This is the reason that cities of the Global South now account for 6 of the top 10 and 13 of the top 20 most populated cities in the world even though, by any measure, their respective formal city economies cannot possibly absorb such populations with all that this implies for the social and environmental conditions in such cities. It is also the reason that in terms of morphology, the best way to describe such cities is that they consist of relatively small islands of downtown gentrification scattered about in a veritable ocean of city slums.

Box 5 The coastal cities of British India

The Portuguese had reached the shores of India and established trading relations by the late fifteenth century. Other European powers soon followed. By the early eighteenth century, the Mughal Empire had weakened considerably both because of internal power struggles and indigenous resistance, and also as a result of the growing power of Europeans, particularly the British, on the subcontinent. By the late 1700s, the British East India Company had effectively become the main power in the area and the last Mughal emperors retained only nominal control over mere slices of their former empire. Like other Europeans, the British initially were attracted by the fine cotton and silk textiles, indigo dye, and handcrafted swords and furniture produced in India. During this period, however, they were beginning to focus more on establishing coastal trading ports to export increasing amounts of bulk raw and semi-processed goods such as cotton, jute, and opium, as well as to import more finished products from their own emerging industrial factories back home.

Unlike the agrarian empire of the Mughals, from its very beginning this new British Empire was based on overseas trade. Centrally located Delhi, which had lost much of its population in the ongoing weakening of Mughal power, was not a priority location for British political control over India. The coasts were much more important for this and so Chennai (then Madras, founded in 1639), Kolkata (Calcutta, 1656), and Mumbai (Bombay, 1652) became British India's major cities. As with the Mughals, each of these cities was centered around a major fortress within the walls of which most of the European settlers, including East India Company executives and later British government officials, lived. The monumental fortress thus, again, physically set off the new conquerors from the conquered in the space of the city.

Like Mughal city rebuilding, the massive monuments of the new British-dominated cities in India were also intended to exhibit overwhelming military police, political, and economic power. In general, the social and built morphologies of these emerging coastal metropolises evolved on the basis of this colonial relationship with the central port and city developed for Europeans, and the indigenous immigrants from the countryside relegated to specific areas – so-called "Black Towns" – outside of these central areas. Hence, these coastal cities represent another example of what have been called the common dualist structure of most formerly imperialized cities of the Global South. Here, however, emphasis is placed on the wider impact upon city development of British imperial intrusion. First, there was a clear reorientation of internal Indian development as, increasingly, resources, people, and the transportation networks they followed were drawn away from the interior of the subcontinent toward rapidly growing coastal port cities. Second, the increasing focus of India's British overlords on raw material cash cropping expanded these activities at the expense of further development of Indian manufacturing. At the same time, Indian consumers of cotton textiles provided a captive market for the growing output of factories in Britain, surely a win–win situation for the British Empire. Finally, the power of the British to impose their will upon city life clearly ensured a pattern of city development that formally enforced discrimination and unevenness in the quality of social and material life within these cities to an extent unheard of even in the most dire city circumstances in Europe.

FURTHER READING

Robert Bruegmann provides a very accessible account of recent city development in his *Sprawl: A Compact History* (Chicago, IL: University of Illinois Press, 2004). The topic of gentrification is treated in breadth and depth in *The Gentrification Reader*, edited by Loretta Lees and others (London: Routledge, 2010). Ananya Roy and Nezar Alsayyad have edited a book, *Urban Informality: Transnational Perspectives from the Middle East, Latin America, and South Asia* (Lanham, MD: Lexington Books, 2004), with chapters dealing with the recent evolution of city labor markets in the Global South. For an early yet still good general account of post-industrialism, H. V. Savitch's *Post-Industrial Cities* (Princeton, NJ: Princeton University Press, 1988) is still to be recommended. The broader historical context of Chinese city history can be found in R. Keith Schoppa's comprehensive *The Columbia Guide to Modern Chinese History* (New York, NY: Columbia University Press, 2000). The plight of U.S. inner cities and the people who live there is described by William J. Wilson's classic *When Work Disappears: The World of the New Urban Poor* (New York, NY: Alfred A. Knopf, 1996). Finally Rajnarayan Chandavarkar rounds out the history of Indian cities in his *History, Culture and the Indian City* (Cambridge: Cambridge University Press, 2009).

CITY POLITICS

In many ways, cities are the very source of politics as a means to manage the new, much more complex social relations that city life necessarily entails. Unlike rural and village life, cities are populated by a much greater diversity of people who are generally not related to each other. City people come into relation with relative strangers on a daily basis and in very close proximity to one another. People with much different belief systems, behavioral norms, and day-to-day rituals must somehow "get along" for the city to even function socially. This new and densely packed civil society is considerably different from rural and village life, where human groups generally manage their relations via extended family ties or via traditions established among a very few extended families. In cities, new rules and regulations must be created in order to manage social relations among a much wider array of people.

The need to maintain social order, then, is the major reason why politics emerges most clearly in cities. The diversity of city politics over human history and in different cultural settings is thus a result of differences in how and by whom such order is maintained. The politics of Ancient/Classical cities, for example, was mostly authoritarian, ruled dictatorially by a small number of hereditary rulers. Indeed, the politics of cities of this time was also generally subordinate to those of the greater empires in which they existed.

City people, in this respect, were not free to manage their own affairs independent of traditional imperial authority, although this authority itself usually was extended from a primary capital city on through secondary and tertiary imperial cities.

By contrast, the politics of mercantile cities was generally more democratic as self-made merchants created self-governed city-states, breaking ties with traditional land-based and hereditary political authorities. These city-states were essentially free to rule themselves. But there were important exceptions to these rather broad characterizations of historical cities. Some Ancient/Classical cities, most spectacularly in Greece, did evolve a more democratic city politics. Indeed, the notion of democratic "citizenship" as mutual obligation of individuals to each other emerged most distinctly in Greek city-states. Greek citizens, for example, were individuals in agreement over social responsibility and rights within the city community. This is an obvious political advance over the notion of traditional political authority over a population of "subjects" with no say in how they are governed. It suggests a new type of politics of individual participation and self-organized community. Indeed, the concept of citizenship arising in early Greek city-states engendered a growing sense of individualism and individual rights and duties that eventually became a major defining characteristic of Western civilization as a whole. Even at its beginnings, then, it is clear that, from the point of view of traditional authority, such a notion of individual participation in community governance is something dangerous, to be avoided at all costs, if possible. That such a notion arose in cities suggests, again, that cities and city people themselves need to be tightly regulated in order to control the growth of such dangerous ideas and social practices. And this is what happened in most civilizations until, as we saw in Chapter 3, the Western Renaissance.

FORMS OF CITY GOVERNANCE

Nevertheless, from its genesis the idea and practice of citizenship has been much more limited than this ideal description suggests. In Ancient/Classical cities, only the most elite males were considered citizens in this respect while the vast majority of other people living in cities, including both elite and non-elite women, were not. These elite males made decisions which essentially served their own interests more than those of the community as a whole. The interests of the majority of the population of the city were not taken into consideration, except when their behavior became dangerous to elite citizens. Granting of full citizenship in the city thus became an instrument of power, control, and ultimate political exclusion.

Indeed, the history of political citizenship has been one of restriction and extension of political power. While originating in ancient Greek city-states, the locus of citizenship eventually shifted to the nation-state, a political form originating, itself, in the West and then spread across the planet by Western imperialism. The key to this history has been who, in fact, has the right to have a say in the politics of the nation? Much like in Ancient/Classical cities, the politics of early nation-states was that in which those white males with property had a right to participate while those white males without property, as well as all women and non-whites, did not. In this context, it was no contradiction for, say, Thomas Jefferson to state that "all men are created equal" with equal rights to liberty and the pursuit of happiness while he, himself, owned African slaves with no such rights. In Jefferson's mind, "men" were naturally "white men with property" who should have full citizen rights in an independent American nation. White men without property, women, and all non-whites were simply not within his horizon of political "citizenship."

The history of city politics in the West has reflected this history of national citizenship. Up until the twentieth century, large groups of people in cities continued to be excluded from community decision-making. It took some time for citizenship to be extended to women, ethnic minorities, and even former slaves, and this extension was not made willingly by the political powers that be. It was, instead, the result of growing demands by excluded groups for full political rights that could not continue to be ignored. Of importance is that these demands by women, the non-propertied classes, and ethnic minorities were voiced loudest in the growing cities of the industrializing Global North. Densely packed in the cities, the unevenly organized and politically excluded classes became most dangerous to the ruling elite. Political concessions were necessary to avoid complete social upheaval, perhaps even city-based social revolution.

The city can thus be considered the site of both the genesis and the extension of citizenship rights and the expansion of political participation. But this has not been a smooth, untroubled history of political enlightenment on the part of elites. Rather, it has been one of violence and social unrest, from the urban revolutions in Europe in mid- to late-1800s, to the globally widespread urban rioting of the 1960s and more recently. It is not surprising, then, that in today's rapidly globalizing world it is in cities where new questions of citizenship have arisen first and been the most publicly contested. As international migration has increased dramatically, the citizenship rights of immigrants, both documented and undocumented, has been greatly debated in recent years. To the extent that citizenship has not been extended, or extended fully, increasingly large numbers of people in the major cities of both the Global North and the Global South are simply excluded from the political community of the nation, as well as of the cities in which they mostly reside. This is hardly a healthy situation for the creation of stable social relations and, in this respect, it represents a probable source of perhaps violent social unrest in cities on a planetary scale in the not too distant future.

REAL CITY POLITICS

So the use of the term "democracy" to describe the politics of mercantile city-states, let alone classical Greek city-states, is a bit of a stretch. The overwhelming majority of city people had no real say in the way they were politically managed in either case. Indeed, this very fact helped to spread Renaissance merchant culture throughout Europe (and eventually the world) as those merchants and other city people shut out of political power and influence in any particular city set off to locate in other cities where such opportunity might be found. In any case, over the course of human history cities have been sometimes quite subordinate to larger political communities such as empires and nation-states or, in fact, more politically independent such as the Greek and Renaissance city-states. Indeed many argue that today, after several centuries of being quite subordinate to nation-state governments, metropolitan areas around the world are again becoming more politically independent (see Chapter 10). The reason why the political status of cities is important is that it determines who has the authority and the means to determine how cities develop and how their internal social order is to be maintained. In short, in politically subordinate cities, the larger political community has the most say in this; in more politically independent cities, it is city-based authorities who have the most say.

There are a number of important implications of this which remain important for the politics of cities to this very day. Before outlining these, however, it is first necessary to think more specifically about the political necessity of maintaining social order within the city. Densely packed, culturally diverse city people need more than mere threat of force or policing to maintain good social relations and to avoid anti-social behavior, although this is necessary, of course. As city populations grow there increasingly needs to be some way to deal with fires and other emergencies, to supply them with water and disperse their wastes, to provide them with decent routes and roads to get around and actively manage this, to make sure of a food supply and its reasonable distribution, and somehow to ensure that housing is adequately supplied in order to avoid both conflict between the homeless and the homed, as well as public health issues related to either cramped living quarters or no quarters at all. In the end, social conflict and other social trouble as well as diseases spread very rapidly in densely packed cities. This means, importantly, that it is in everyone's interests, the rich as well as the poor, to avoid such troubles as much as possible.

Dwelling for a moment on this last idea already points to a major source of problems for today's cities, as more and more of the relatively rich have abandoned cities altogether. In the present context, however, the point is that maintaining social order in the city is a relatively enormous task and, importantly, one that is ultimately quite costly. The enormity of the task suggests the very necessity of creating formal institutions of governance. The expense of the task suggests, in turn, the necessity of some source of public revenue to complete it. City-based public figures must then be designated to manage the reproduction of the social order, and taxes for this purpose must be raised. And here is where the political status of cities becomes important. Theoretically, more subordinate cities can rely on a much bigger pot of imperial or nation-state tax revenue for this purpose, if proper policies for this are put in place. More politically independent cities have to rely on their own internally generated resources.

The continuing fate of cities therefore depends substantially on the greater political context in which they develop. It was the case, for example, that politically independent Renaissance merchant city-states were able to maintain social order on the basis of internal taxation because ever-growing trade was rendering much wealth in such cities. More even city development took place within strong nation-states after the seventeenth century in Europe largely because of the more even sharing of national fiscal resources among them. Indeed, even in today's world of globalization and what some call the weakening of nation-states, cities in European countries with a tradition of such national revenue-sharing still enjoy a relatively even development pattern. But, as detailed below, dominant contemporary trends suggest that even this European difference may be changing toward a more American model of independent, fend-for-yourself city politics and development not unlike that of the early Renaissance cities. Unfortunately, however, unlike the latter, most U.S. cities do not enclose sources of ever-growing wealth anymore as a handy means to fend for themselves.

POLITICS WITHIN THE CITY

While the political context of cities is thus important, city politics is really about what goes on within cities. Until this point, in other words, cities have been described as if they were really political actors themselves instead of merely places comprised of many different political actors. The politics of maintaining social order in cities is, instead, more about the use and abuse of differential power within the city. In this respect, it is important to note that the extent and distribution of the public services of social order listed in the last section is determined by those most powerful. In imperial cities, these would be imperial agents who would make their decisions based on imperial interests in the city. In post-Renaissance cities, these would be those most wealthy. The key is that the service needs of these most powerful city people were what were taken care of first, while those of other city people were relegated to secondary or tertiary status if they were accorded any status at all. So, as described in previous chapters, the most powerful resided in the center of the city and the less powerful in ring after ring of the city toward the outskirts. In turn, public services were most dense and efficient in city centers and less and less so as one moved out the spatial scale of the city. Such a pattern of the distribution of public services can still be seen today in most post-colonial cities of the Global South.

The Industrial Revolution changed this pattern rather dramatically. The centers of industrializing cities soon became overfull of much less wealthy, low-skilled people and massive factories of divided labor took the place of merchant warehouses and artisanal crafts workshops. This process put enormous pressure on the public services designed for a much different city population. Everything from police and emergency services, to accessible streets, to water supply and waste disposal, to, of course, housing facilities were now sorely inadequate in older established cities such as London and sorely needed in newer, fully industrializing cities such as Manchester. The maintenance of social order became, as a result, that much more difficult, rendering cities, in turn, increasingly dangerous places to reside for everyone, particularly, of course, the relatively wealthy.

So, with industrialism came the growing abandonment of central cities by the very people with the most means to help fund the very public services now in increasing demand. Indeed, from this point in time until today's metropolitan world of sprawling suburbs and exurbs, this became a continuing theme in city politics. Here, however, the focus is on this now growing need for public services in cities increasingly populated by the least wealthy. Again, it is in every city person's interest to maintain public safety and public health because of the close proximity of everyone's living and working quarters. Particularly in the early phases of industrialization when even those who moved to the suburbs still worked in the city, this truly public interest in the maintenance of social order remained relatively high. As industrializing cities continued to overfill with growing numbers of working-class people and their needs, it did not take too long before it was clear, even to traditional city elite, that something had to be done to at least minimally meet such needs.

But this realization on the part of elite city dwellers and other traditional authorities did not come easily. It took much prodding by social activists in city after industrial city hoping to "reform" the way in which the new city people were provided with housing, freshwater, and adequate health and public safety. It also took open conflict in the form of urban riots and revolts that have erupted regularly, from the mid-nineteenth on into the early twenty-first century. Of most importance is that these urban reform movements gave increasing public voice to the growing numbers of working- and other lower-class city people. As a result, the reproduction of social order in the city became increasingly a publicly negotiated affair among all classes, not just that of elite authority. The masses, in other words, were increasingly given a much louder, more formal political voice in the city and, because of this, they found in city government a new source of political empowerment altogether.

REFORMING CITY POLITICS

But, again, these were hard-won political rights. City reform movements faced a steep uphill battle. Post-Renaissance cities have developed on the basis of private business run entirely for, and on the basis of, private profit. From the mercantile period on through the post-industrial period, it essentially has been private business decisions that have determined how most cities have fared. Yet, surely, reforming public services for the city masses had to be paid for somehow and, in the early days of reform, this meant mostly by the small numbers of city-based economic elite. How this worked itself out differed from city to city; but, at this level of generality, there are three things to notice about this issue of importance for the present account. First, this new public demand put on the economic elite gave added push for them to move out of the city altogether. In fact, it still does at the beginning of the twenty-first century as most wealthy suburbanites and exurbanites attempt to avoid political annexation by central cities to keep public tax revenue for their own use closer to home.

Second, the need to gather and maintain higher levels of tax revenue to implement needed reforms in the city meant that the business people who did remain in the city had to be accommodated in significant ways for them to stay in the city and pay taxes. In other words, while the majority of city masses acquired increasing political voice and power in city politics, they still had to cater to private business needs in order to ensure adequate tax revenues for public reforms. City politics thereby became of necessity one of a coalition of public- and private-sector actors working together for the benefit of the city as a whole. As described below, cities in this way became increasingly governed by a **political regime** of politicians and major private-sector actors.

Finally, the necessity of increasing public revenue for needed public reforms in the city rendered the wider context of city politics that much more important. In other words, more politically subordinate cities came to rely even more on fiscal redistribution from the nation-state for their continuing development. Most nation-state political authorities, in turn, were largely agreeable to this because, as time went on, city populations increasingly made up the majority of national populations. Cities also became the main economic engines of national economies. Finally, cities were sites where major social disruptions usually emerged, from outright revolutionary movements to more localized, yet quite socially disruptive, city riots. Keeping city populations largely mollified and orderly, then, was important for the nation as a whole.

The extent of nation-state involvement in city politics and development varies from country to country. Nevertheless, a broad characterization can be made. In Europe, for example, nation-state governments continue to play a large role in city affairs. As discussed again in Chapter 9, this is manifested not only in the way in which public services are maintained, but also in the more compact built and social morphology of cities. Nation-states in post-colonial countries also continue to play a large role in city affairs largely as a legacy of their dependent imperialized development where political power emanated singularly from the imposed capital city. In the United States, however, the nation-state generally has played a much lesser role in city politics and development. The reason for this is related to the more decentralized political system created by the country's founders on the basis of a deep distrust of too much centralized power. This has meant that, unlike cities in Europe and elsewhere, U.S. cities have generally had to fend for themselves in order to develop successfully. While there have been times of more national government involvement, particularly during the Great Depression and in the Great Society programs of the 1960s, for the most part U.S. cities have been on their own in this regard. This has meant that city politics in the United States always has been about, primarily, securing an adequate tax base for funding necessary and discretionary public services. City politics always has been, by necessity, more entrepreneurial and less managerial in this regard. This is important to note because if, indeed, nation-state power all over is being diminished rapidly by the forces of globalization today, as many people argue, then even European and post-Colonial cities may be left to fend more for themselves like U.S. cities. If this is the case, then the historical and contemporary experience of U.S. cities may be a harbinger for cities across the planet in the twenty-first century, as noted in Chapter 1.

THE POLITICAL CONTEXT OF CITIES IN THE TIME OF FORDISM

Yet, the political context of U.S. cities has not been as cut and dried as this rather abstract account of general trends suggests. Nation-state policies, particularly after World War II, had rather enormous, if largely unanticipated, effects on cities. As described in Chapter 4, the new larger role of the national government in the post-war U.S. economy was the result of a conscious decision made by policy-makers to avoid another economic depression like that of the 1930s. Again, the two major national policy initiatives which had the most initial impact upon cities were the nationally backed mortgage assistance program for returning war veterans and the major interstate highway acts. Both of these policies of the national government greatly accelerated the suburbanization of both jobs and middle- and upper-class people and therefore much needed potential tax revenue. This put a heavier tax burden on those who remained in the city, most of whom were in no position to pay. Such a burden also led to more suburbanization on the part of those, individuals and businesses, who were capable of paying higher taxes, thereby fueling a vicious cycle of economic decline and dereliction of public services in central cities.

These national policies thus played a very large role in creating the pattern of contemporary city development in the United States regardless of the fact that they were not conceived as policies specifically for cities. Moreover, these specific policies, if formulated differently, could have had a much different impact upon city development. For example, if the housing policy for war veterans had been less biased toward the ownership of single-family homes and more biased toward more densely populated, multifamily, city-based housing or even toward subsidizing the rental housing market, U.S. cities would have developed much differently. Similarly, if national transportation policies had been less biased toward the private ownership of automobiles and road development and more focused on modes and networks of public transportation for the masses, cities would have been affected much differently. Indeed, these alternative national policies essentially were enacted in European countries as they rebuilt their cities after the devastation of the war effort, and this still explains the distinctly different development trajectories of European cities *vis-à-vis* those of the United States.

The specific reasons for these differences in the bias of national policies range from the nature of the respective political processes, to the greater territorial space of the United States, to the much more decentralized tradition of U.S. politics, on to the much stronger bias in the United States toward relying on the private sector to determine the best public policy to pursue. Wherever the emphasis is placed, the point is that these are political decisions that were made at the national level, for whatever reason, that greatly affected what happened in cities. In the U.S. context, these policies greatly accelerated a process in which the economies of central cities declined; increasingly, the great majority of the residential population in inner cities came to consist of people with the most need for public services. And given the racial history of the country as well as active discrimination in suburban housing markets, this majority inner-city population increasingly came to consist of African-Americans and other people of color.

So the central areas of city after city in the United States became increasingly derelict as factories and warehouses were abandoned and landlord after landlord walked away from their properties for lack of even a viable rental market. The dwindling tax base of cities simply exacerbated this situation as schools could not adequately be funded any more than properties could be properly police-guarded or generally kept from falling into complete disrepair. By the 1960s, indeed, the inner cities of major industrial cities such as Detroit, Chicago, and Los Angeles were bleak places of deteriorating built environments and increasingly frustrated and even destitute inner-city dwellers densely packed in mostly substandard housing. In this context, all that was needed was a spark to set off an explosion of pent-up anger among inner-city residents with little hope and large feelings of economic neglect and social discrimination.

THE REALITY OF URBAN RENEWAL POLICIES

In this context, another important set of national policies which became known as **urban renewal** needs to be discussed, particularly because they were more specifically targeted at cities. As noted, the U.S. federal government was interested in proper housing for returning veterans and this included doing something about the increasingly crowded substandard housing in inner cities. Particular focus was put on inner-city neighborhoods in the direst social and built condition, as well as on districts of abandoned and derelict factories and warehouses. Beginning with the Housing Act of 1949 and gathering steam with a further Housing Act of 1954, this project of urban renewal was based on the use of federal funds to buy up such properties using the power of **eminent domain**, or the legal taking of private property with compensation for official government business.

More properly known as slum clearance, the policy of urban renewal put large tracts of particularly inner-city and formerly industrial land into the hands of city governments. Whole neighborhoods were lost in the process and all the housing, substandard or not, that this entailed. At the same time, such slum clearing housing policies became combined with the federally funded interstate highway initiative which greatly accelerated the demolition of whole neighborhoods and business districts of the already poor and mostly people of color. New intra-urban highways plowed right through cities and directly through such neighborhoods of least possible social and political power, thereby, from the point of view of the federal government and most city authorities, killing two birds with one stone: highway development and slum clearance.

Such federally sponsored urban renewal or, as some called it at the time, "urban removal," brought with it four main results for cities. First, as whole neighborhoods were demolished, the remaining housing in inner-city areas became ever-more crowded and expensive. Second, the publicly funded replacement housing for displaced residents of such "renewed" areas tended to concentrate the most poverty stricken in densely populated, large-scale, multifamily public housing projects. Such housing projects were also frequently built alongside new intra-urban highways, which imposed an actual physical barrier between such projects and the more affluent areas of the city, spatially concentrating the poor even more. Third, what few jobs remained in inner cities that matched the skills of most inner-city residents became even fewer as business districts in poor neighborhoods were cleared and as the new intra-urban highways made it even easier for businesses to move out of the city. Finally, because federal funding was extended only for buying and not developing the city land bought, city authorities were not required to replace housing in the same amount that was lost nor were they required to replace such housing at the same level of affordability for those residents displaced.

All of these results of urban renewal simply exacerbated the plight of inner-city residents already adversely affected by the suburbanization of jobs and the middle and upper classes. Indeed, city development as a whole was adversely affected as renewal and highway development accelerated the suburbanization process, in general. Because of this, most city politicians sought to redevelop the city land they acquired in order to cater to the needs of more affluent people and businesses. This is the early source of the ongoing trend toward the gentrification of central city housing and business so prevalent in city after city today. Facilitating the conversion of central or inner-city housing from low-rent to high-rent housing, as well as the creation of publicly funded city amenities, from parks to river walks to convention centers to museums and sports stadia, etc., became increasingly popular with city politicians in order to attract more affluent taxpayers back to the city. Such projects, particularly undertaken by the private sector, could be highly profitable themselves, in fact, because they were being built on relatively prime central city land that was acquired very cheaply as a result of urban renewal policies.

GROWING SOCIAL RESTLESSNESS IN THE CITY

National urban renewal policies thus helped to accelerate the creation of the now familiar development pattern of post-industrial cities in the United States, with gentrified, affluent CBDs and adjacent housing surrounded by sites and whole neighborhoods of concentrated poverty and dereliction. The more immediate impact of urban renewal, however, was that it greatly exacerbated the economic and housing situation of inner-city residents, particularly those African-American and other minorities actively discriminated against in both the labor and housing markets. Again, as early as the 1960s, the combined effect of suburbanization and urban renewal had created an extremely tense social environment in most inner cities which eventually came to a head in 1965 in the Watts district of Los Angeles when a full-scale riot took place over a good

30 hours leaving over 30 dead and involving some 4,000 arrests. This riot marked the beginning of a series of inner-city riots over the next few years involving Chicago and Cleveland in 1966 and Newark and Detroit in 1967. The riots in Newark and Detroit were particularly large and took place over a number of days, leading to literally thousands of arrests.

The immediate spark for such riots usually involved allegations of abuse and oppression on the part of largely white city police forces brought to bear on neighborhood populations of color. But the real underlying roots of the unrest were to be found in the very precarious economic situation of inner cities just discussed. Of importance, these types of outbursts on the part of inner-city residents have continued in the United States for much the same reason. There was a riot in Miami in 1980, for example, that resulted in over 1,000 arrests and 18 deaths. Even more recently, Los Angeles erupted again in a 1992 riot that resulted in as many as 58 deaths and about 17,000 arrests. While the specifics of each riot differ in several respects, the main cause does not. There continues to be a major spatial mismatch between skills and jobs that does nothing but fuel the underlying frustration that periodically vents itself in inner cities. Indeed, it is just this sort of mismatch, combined with the decline in social services as a result of today's neoliberal city politics, that has fueled the most recent riots in the industrial suburbs of Paris and Lyon and, even more recently, those of London, Manchester, and Liverpool.

The result of such unrest in the city essentially has been threefold. First, national attention is refocused on inner cities and new policies are put in place or old ones reinvigorated in order to address the situation, even though policies such as urban renewal actually served only to exacerbate the situation. Second, however, the kind of attention that is newly focused on cities by non-policy-makers tends to render the situation worse by prompting more and more businesses and people to want to avoid central city locations altogether. In the popular mind, the view of cities as dangerous places in which to live and conduct business simply becomes reinforced. Third, when faced with the threat or reality of major social unrest, most city politicians and their allies have sought to disperse it by facilitating further gentrification and urban renewal in order to break up concentrations of poverty, especially immediately adjacent to the CBD. They also have sought to reinforce police presence and coverage and allowed the privatization of formerly public space similarly policed by private forces, as discussed further in Chapter 9. In the end, these results of social unrest in the city have generally not solved the underlying problems that give rise to such. In many cases, indeed, these problems have been exacerbated or, in somewhat better circumstances, simply displaced to other parts of the city.

CONTINUING THE PRACTICE OF POLITICS IN CITIES

There is a wide array of different forms of city political relations from what are called strong mayor systems to city council mediation to regional governance umbrellas and so on. In general terms, the actual politics conducted within these institutional forms can be considered as falling somewhere on the continuum of what have been called more **elitist** or more **pluralist** processes. More elitist politics is where a relatively small group of people make most of the political decisions with very little direct input from the greater city population, save for, in most cases, periodic formal votes. In more pluralist processes, more input is elicited from a wider array of city people via, for example, larger city councils, the consultation of citizen boards of representation, or more regular plebiscite of the general city population.

Similarly, there is a continuum of political styles within cities from what can be considered a more managerial style to a more entrepreneurial one. Generally, cities which have a more subordinate political role within a larger political arena, like those of Europe and of post-colonial countries, tend to adopt a more managerial style as a result of having to ensure the distribution and effective management of funds collected in this larger political arena. Indeed, how such funds are used is mostly already determined at a higher level of politics so that city politicians simply have to ensure that this gets done properly. On the other hand, cities which are less politically subordinate to nation-state or higher levels of politics, like those of the United States, have had, by necessity, to be more entrepreneurial in the sense of ensuring their own development with their own resources. The fact that many are arguing now that all cities increasingly have to go down this road to a more entrepreneurial politics because of globalization suggests a closer look at such a city politics is necessary.

A first stab at this is to recognize, again, that cities do not act themselves, but that actors within cities do. Who, then, are the major actors in the politics of the city? In more managerial cities, these can be political appointees. In more entrepreneurial cities, major actors are usually determined in place and are usually strongly committed to the city itself. These are wide generalities, of course, but they make sense since in the latter cities politicians and their allies need to be, in the first instance, economic **boosters** for their cities in order to ensure continuing development. This has led many commentators to suggest that the politics of entrepreneurial cities really goes beyond what happens in the formal institutions of government, whether these are dominated by strong mayors or city councils or whatever. For such commentators, political power in the city is rather diffused among both politicians and major private-sector actors who form a relatively coordinated regime of political authority and, then, city **governance**.

The idea behind this conception of regime politics is that, in market-based societies, political power does not just reside with politicians but also with major businesses and other private-sector actors with a major stake in the development of particular cities. This is true even in more managerial cities because city development as a whole still necessitates economic success. It is most apparent in entrepreneurial cities, however, because such cities have to largely fend for themselves in this respect. U.S. cities, for example, have always been governed by ruling public–private coalitions of powerful city people who have acted as the main boosters for city development regardless of whether or not such people were chosen by the greater city population to do so. Indeed, in most cases in the past such a coalition has consisted of a close-knit, relatively small group of powerful people who created city policies to fit their own needs first, whether these were to stay in formal political office (and sometimes gaining riches thereof) or in attracting more business either to themselves or as a result of new additions to the city economy. In the past, then, most city politics in the United States took the form of a relatively closed and closely focused economic **growth machine** to the neglect of other social needs of the city population. The prevailing belief behind such public–private political coalitions is that once the entire city economy is growing successfully, the proceeds of such success will trickle down to meet the needs of the greater city population. In fact, this is fast becoming the prevailing belief in the increasingly neoliberal politics of today's globalizing cities.

REGIME POLITICS IN THE CITY

The concept of city political regimes thereby underscores this process of public–private coalition-building in city politics. As cities entered the twentieth century, such coalitions in the United States were forced to incorporate more and more powerful – at least in members – citizen activist groups pushing for things such as housing reform, more adequate provision of health and safety services, and other things for the benefit of the non-economic elite. So regime theorists have categorized some regimes as more elitist or more progressive over time depending upon the kinds of groups that were included in such governing coalitions and, then, the nature of the policies that were formulated, whether singular elite focused or more socially inclusive.

Regime theory also makes it clearer who is most likely to participate in the governance of cities and at what level of participation. In general, it will be those actors with the most invested in a particular city, whether as businesses or as permanent residents or, indeed, as both. In early growth machine politics the most powerful members included hometown newspaper owners, representatives of locally based banks and insurance companies, big department store owners, and major real estate owners and developers. During later periods, as noted above, regime membership in many cities came to include neighborhood groups and social activist groups seeking local reform. The point is that these actors have the most at stake in the continuing development of "their" city.

While seemingly obvious, the importance of this view of city politics is that it suggests that such stakes may be both stronger and weaker and, indeed, may change over time depending upon the nature of the attachment of businesses and individuals to particular places. For example, in today's globalizing world, many hometown newspapers, banks and insurance companies, and department stores have been acquired by national and even international corporations. Similarly, national and international development companies have acquired local land as investment property and many upscale residents of cities are new to the place or soon might be transferred elsewhere for job-related reasons. Because of this, the allegiance of such city-based actors to any one city is increasingly tenuous at best. As a result, city-based growth coalitions today are both less cohesive and yet need to be extra vigilant to attract and, indeed, keep such businesses and people in any particular city. They must govern cities with the interest of these groups foremost in mind, which, in turn, has resulted in two things. First, city growth coalitions are increasingly competing with each other with the same sort of publicly expensive tax and subsidy and city amenity policies. Second, again, the needs and desires of less powerful people of the city tend to be neglected in the pursuit of such economic growth in the belief that these will be taken care of once such growth occurs.

And so the contrast between the economic condition and prospects of inner-city neighborhoods and the gleaming central cities and CBDs of globalizing cities is becoming even greater. This has led some to argue that a resultant **dual city** structure now characterizes most major cities across the planet, whether in the richer Global North or the poorer Global South. Others argue, however, that this polarization of wealth and poverty in the built and social environments of many cities today is less a simple duality and more a patchwork mosaic of metropolitan bits of wealth and poverty in spatial juxtaposition, yet with, unfortunately, very little direct relationship, either economically, socially, or culturally.

ECONOMIC CRISIS AND THE EMERGING NEOLIBERAL CITY

As discussed in Chapter 4, the continuing post-World War II economic growth of major industrialized nations came to a halt with the crises of the 1970s, brought on by increasing international competition and the drastic rise in energy costs. This brought forth the perceived need to restructure the way in which things were produced, what was produced, and closer consideration of where such production should take place to remain most competitive, as we have seen. Of importance for city politics, this new post-Fordist mentality carried directly over into the political realm as the belief became increasingly that the plight of cities was largely due to ineffective political leadership and regulatory red tape. In other words, city-based political institutions came to be considered overly bureaucratic, cumbersome, expensive, and, as such, largely unable to meet the needs of a new economic reality. This perception was reinforced by the growing perception that large-scale federal government policies such as urban renewal and the so-called War on Poverty had not solved city problems but, in fact, had made them worse. In short, the growing perception was that big city government, like big federal government, was more a cause of the economic crisis of cities than a solution.

Not surprisingly, the solution generally found was to downsize big bureaucratic government in the same manner that private corporations had to become leaner and meaner and more flexible to deal with the post-Fordist economic crisis situation. Government itself, whether at the national level or that of the city, was increasingly seen as the problem. In fact, this is now the common understanding of the situation even after the major global recession of 2008, so it does not need much more elaboration here except to note that this understanding has come to be called **neoliberalism**, which signifies a new type of *laissez-faire* approach to market processes.

This growing neoliberal persuasion led to drastic cutbacks in federal government support for general social welfare policies, as well as policy initiatives more directly related to city development. Such cutbacks merely reinforced the fend-for-yourself development context of U.S. cities in which only local sources of revenue were available for this purpose. This exacerbated the already increasing **fiscal squeeze** felt by city after city as a result of de-industrialization and the suburbanization of business and the upper and middle classes. City authorities had to assume a heightened entrepreneurial role in this context which intensified, in turn, the competition among cities to attract or retain business and upscale residents in central cities. Indeed, given that property taxes are the most important source of city tax revenue in the United States, this heightened entrepreneurialism of city policy necessarily had to be focused on the wealthy firms and individuals who would increase the value of city property via gentrification.

In general, this new context of neoliberal city politics marked a significant shift in the understanding of the role of government. From social distributive and general social welfare policies, the role of city government was increasingly considered to be that of providing and strengthening measures to ensure the economic competitiveness of cities. Policy decisions affecting the city were, and continue to be, made on very narrow economic cost-benefit calculations rather than on much broader, more traditional political criteria of community service and social welfare for the entire population of the city. In the best of all possible worlds, the tax proceeds of economic growth engendered by these new entrepreneurial city policies would, indeed, be spread to the entire population if only in a trickle-down fashion. However, in the reality of now global competition to attract and retain the national and international gentry, most of this revenue has been plowed right back into the sorts of city projects created to attract such actors. Increasing amounts of the built and social morphology of central cities have thus become fully **imagineered** to match the perceived desires of global elite actors, from wholly reconstructed and sanitized waterfronts and river walks, to massive hotel-office-convention centers, etc., to the "cleaning up" or shooing away of undesirables, such as the homeless and other "vagrants," as well as much heightened police surveillance.

Indeed, the necessity of relying relatively solely on local revenue for city development has also led to the search for other possible sources of tax dollars. As a result, there has been a proliferation of new or increased rate revenue boosters such as user charges (road tolls, park and museum entry fees, etc.), local sales taxes, and taxes on such localized activity as gambling and hotel stays. Overall, such new or increased fees have made city living more and more expensive and then less and less feasible for those least able to afford the new costs involved.

Finally, given this new fiscal context of neoliberalism, the other major political movement within cities has been toward privatizing city services by outsourcing formerly public functions from construction to the maintenance of city vehicles, to garbage collection, to public housing and sports stadia management, to aspects of policing and public health, etc., to private firms in the search for perceived cost savings and more efficient delivery. There also has been a general growth in **public–private partnerships**, particularly for economic development in the form of publicly subsidized city-based development corporations intended to facilitate the private redevelopment of city property. In addition, publicly subsidized **enterprise**

zones within the city have been created to entice private firm relocation via tax breaks and other incentives. Even in the realm of education, a major public expense for city governments, partnerships with the private sector have been created in the form of so-called charter schools supposedly supplying better educational opportunity in a more economically efficient manner.

In this respect, even more than the city political regimes made up of coalitions of major public- and private-sector actors, this new form of increasingly privatized city governance has become the norm not just in the United States but in most globalizing cities of today as a result of neoliberal policies. The consciously planned and increasing gentrification of the built and social environments of central cities around the world has, indeed, succeeded in creating places of much polarized wealth and poverty. Combined with the ongoing suburbanization, or gated isolation, of relatively wealthy city people, this polarization within cities has become quite dramatic. Because of this it has become much more difficult for citizen activist groups to form, let alone achieve success, in resistance to further gentrification and social displacement within central cities. More and more political regimes in cities have become dominated by those most sympathetic to the needs of, or actually allied with, the global economic elite. In this respect, neoliberal politics in today's cities most resemble the solidly elite economic growth machines of cities in the past. Now, however, such city-based growth machine politics exists not just in the United States but, increasingly, worldwide.

Box 6 From drug war to unilateral treaty ports in China

China can be said to have had the longest continuous history of more or less unified city-sourced civilization of any on Earth. The depth of civilization that such a level of continuous city life represents left foreigners who visited China, such as Marco Polo, simply astonished. Yet, the British who began to appear off the coast of China during the eighteenth and nineteenth centuries were soon to change this perception, though, initially, even they could not persuade Chinese authorities to engage in open trade relations with them. Indeed, the Chinese relegated all trade with these persistent new foreigners to specific cities on the coast, first Guangzhou (Canton) and later others, where such foreigners could be heavily monitored.

The British, however, eventually found a way around this tight Chinese regulation. Having already conquered India, they established a cash crop – opium – that could be produced there which had a growing market in China. This was another win–win proposition for the British Empire. Even though the Chinese authorities from the very outset of this drug trade had attempted to limit the inflow of opium, the British began to supply Chinese smugglers on an ever-increasing level. In fact, the amount of opium reaching China became so great that, in 1839, the emperor sent a government agent, Lin Zexu, to put a stop to the illicit trade altogether. He did, eventually dumping a whole British merchant ship's cargo of opium into the sea as an exclamation point.

The British considered this an act of aggression and immediately sent the powerful British navy to fight a war *for*, instead of *on*, drugs. Once the British navy steam ships were on the scene it did not take them too long to dispatch any Chinese resistance, given that China had no real navy and that their main defenses had always been directed toward the traditionally hostile northeastern land frontier. The culminating one-sided Treaty of Nanjing (1842) and subsequent Treaty of Tianjin (1858) not only ensured the continuance of this very profitable drug trade, it also established British "rights" to establish permanent trading establishments in many other port and riverine cities in addition to Guangzhou, such as Xiamen, Fuzhou, Ningbo, and Shanghai. These "treaty ports" thus forcibly opened China to overseas trade with Europeans and, as time went on, virtually all major European imperial powers, and eventually even Japan and the United States – with Americans trading great quantities of Turkish-grown opium – had established their own treaty port trading posts on Chinese territory. These powers thereby established their own sovereign pieces of city territory within China where foreign customs, laws, and business practices prevailed.

There is much more to this story, of course. Here the focus is two-fold. First, this foreign intrusion on the sovereign territory of China shifted both the locational and functional aspects of Chinese cities. Growing overseas trade meant that the treaty ports were bustling with commercial activity in which even many Chinese took part. The treaty ports rapidly grew in population and economic power as a result, as railroads were built from the interior to the coast, new commercial banking was established, and even city-based industry began to emerge. Second, most Chinese were perfectly aware that the so-called "treaties" that established foreigner rights to these ports were entirely one sided, with China gaining nothing from their concessions of territorial sovereignty. Indeed, this simply reinforced the prevailing Chinese belief in the morally bankrupt barbarity of these foreign intruders.

FURTHER READING

A recent account of city policy issues in a range of contemporary cities can be found in Neil Brenner and Nik Theodore's edited volume *Spaces of Neoliberalism: Urban Restructuring in North America and Western Europe* (Malden, MA: Blackwell Publishers, 2002). Andrew Jonas and David Wilson provide a fine historical and contemporary exploration of city growth policies in their edited volume *The Urban Growth Machine: Critical Perspectives Two Decades Later* (Albany, NY: SUNY Press, 1999). A comprehensive look at new post-World War II urban circumstances is found in Neil Smith's *The New Urban Frontier: Gentrification and the Revanchist City* (London: Routledge, 1996) and Gregory Squires's edited book *Unequal Partnerships* (New Brunswick, NJ: Rutgers University Press, 1989). Ronald G. Knapp provides a historical overview of *China's Walled Cities* (New York: Oxford University Press, 2000) and the broader historical context of Chinese city history can be found in R. Keith Schoppa's comprehensive *The Columbia Guide to Modern Chinese History* (New York: Columbia University Press, 2000). Finally, Elizabeth Strom and John Mollenkopf provide a broad yet in-depth overview of city politics and issues thereof in their edited volume *The Urban Politics Reader* (New York, NY: Routledge, 2007).

7

CITY CULTURE

A running theme throughout this book is that cities represent unique sorts of places of human interaction quite different from rural and small village settings. With many relatively diverse individuals in close spatial proximity to one another on small territories, cities literally throw people into each other on a regular day-to-day basis. For this reason, life in cities engenders, by necessity, new types of human relations, or city cultures, as individuals negotiate their relations with others who they may or may not know very well. At the same time, issues arising from human relations, in general, like poverty, crime, and other socially deviant behavior, social and ethnic group rivalry, etc., are magnified in city settings as they become obvious much quicker as a result of the rapidity by which such issues affect so many people in such close proximity at once. Densely populated cities in this respect seem to make such issues more intense or, at least, more intensely felt and therefore more obvious. Indeed, this is why many still consider such social issues as those related to city life itself instead of issues of human relations, in general, which just happen to be much more noticeable in densely crowded cities.

Similarly, as described with regard to human population and ecological problems in Chapters 3 and 8, respectively, city life itself can make it appear that more general human problems exist there when they, in fact, may not. The rapid cramming of more and more destitute, or otherwise less fortunate, people into the streets of London, Manchester, New Delhi, and other rapidly growing megacities of the Global South, for example, has made it appear to many, both in the past and currently, that the world as a whole is overpopulating. The clearly observable, largely adverse effects on non-human nature of dense populations of humans in cities suggest, in turn, that humans are in the process of trashing the planet, in general. Yet, the extent to which both of these perceptions may be the reality of the situation is quite debatable. The point here is that city life essentially magnifies such issues by rendering them that much more visible.

More generally, from the very beginning of city life humans have thought of city culture as unique in very important ways. Cities were early sites of social power, as we have seen, but they were also places where new human beings came to be formed, whether as corrupters or generators of new cultural norms and values. In religion after religion in human history, for example, city life has been considered as a source of corruption of individuals and groups, leading them down paths of overly prideful worship of human powers as opposed to more rightful or traditional otherworldly power. This makes sense, particularly given the way in which the very existence of cities represents that much more human power over non-human nature. To recover more authentic reverence or respect for a greater power, then, many religions have considered it necessary to exile oneself to the countryside, the wilderness, or otherwise one's own solitary contemplation. Indeed, this quite generalized notion of the human hubris represented by city life led many to consider that the only salvation available for humans was to actively destroy cities altogether from Jericho to Sodom and Gomorrah.

This overly generalized account suggests that anti-city sentiments have a very long past, now perhaps deeply engrained in the human psyche. But the flip side of such sentiments, of course, is that which celebrates the great cultural innovations that have arisen first in cities as the basis of civilization, in general. In this sense, the perception of city culture has always been ambiguous, shading toward the bad or the good depending upon the time and place. It is also the result of who, in fact, is doing the perceiving. As we have seen, city populations have always included a number of non-elites who, at different times, have been considered dangerous to the well-being of traditional city elite. To the latter, cities may, indeed, become dangerous places from time to time and therefore escape comes to be considered absolutely necessary.

Such ambiguity in the perception of city culture continues today. This is important to state so directly here because what one makes of city culture determines whether or not one supports their continuing development or even existence – that is, does it really matter that central cities largely have deteriorated and been abandoned as whole regions have become metropolitan with rather autonomous citified bits interspersed with ever-sprawling semi-rural suburbs and exurbs? In this context, is true city life worth saving at all? Similarly, what one makes of city life determines how one believes social issues within cities should be handled. Is the solution to inner-city social conflict one of more general social welfare or one that just needs better policing as suggested by the current national government of the United Kingdom in the face of very recent city rioting?

THE NEW CITY CULTURE OF INDUSTRIALISM

A wider consideration of the ambiguity with which city culture has been considered over a longer human history is certainly beyond the scope of this book. Here, focus will be first on the rise of merchant city culture and then, more thoroughly, on the rise of industrial city culture. In terms of the former, it is clear that city merchants and those they supported came to regard their independence from traditional land-based authority as a breakthrough to a more progressive, more humane culture. Similarly, as recounted in Chapter 2, traditional land-based elite saw these new cities mostly as dangerous places filled with dangerous people. But within these cities, social life remained highly regulated, particularly that of the lower classes. City people knew their place and kept to it, both in terms of what they did in the city and where they lived. Indeed, this place was even marked by what clothes they were allowed to wear. In this respect, the traditional, hierarchical social norms of the countryside were largely reconstructed over time in merchant cities. And this is another reason why less elite merchants increasingly found it necessary to move to other cities as they faced fewer and fewer opportunities for advancement in longer-established merchant cities.

Traditional, hierarchical social norms, however, were burst quite asunder by industrialization. Changes associated with industrialism came very rapidly. Indeed, they came much too rapidly for traditional social regulation to control. As we have seen, agricultural innovations and management sent waves and waves of people, by necessity, to cities in a now continual process of rural-to-urban migration. Innovations in production and transportation came quickly and artificial clock time and street lighting replaced natural rhythms of work and day time. In short, unlike the already innovative changes that took place in the merchant cities of the Western Renaissance, changes of the industrial period seemed much more rapid and ubiquitous and, indeed, out of control, particularly in the rapidly filling central districts of industrializing cities.

For this reason, there arose much contemporary commentary on the social implications of the growth of industrializing cities that many argue remains relevant today. In the later part of the nineteenth century, for example, social commentators, particularly in newly industrializing countries such as Germany and France, began to consider industrial city life as, in fact, a new human culture in the making. These commentators differed in the specifics of their respective analyses of this new city culture; but they more or less circled around a central theme most clearly enunciated by the German social historian Ferdinand Tönnies (1855 to 1935). In Tönnies's time, Germany was very rapidly industrializing and urbanizing according to conscious government planning because of the perceived need to catch up with more industrialized European countries. Germans, in increasing number, were quite literally being thrown into factories in rapidly growing industrial cities.

LOSING THE MORAL ORDER OF "COMMUNITY"

Tönnies argued that this rapid urbanization was pushing many Germans into a whole new way of life without adequate cultural preparation. Most of these new city people had been socialized in the countryside or in small rural villages. It was there that they came to understand the norms and values of their society. Now, however, they were being thrown together with more and more people whom they did not know and whose norms and values were quite unfamiliar, if known at all. Specifically, Tönnies argued that what he called the "moral order" of the countryside was no longer feasible in the industrial city. In small village life, for example, everyone knows each other and they all work in similar jobs in similar ways. Everyone knows their role and responsibilities and everyone knows each other's. Social relationships are longstanding, deep, and based upon very familiar, if not actually family, ties that bind. Social control is exerted by the largely informal discipline of families and near neighbors, making sure that everyone keeps to their place to ensure peaceful, steady, social stability. In short, in village life in the countryside, what is permissible and proper social behavior is made more or less fully known to all as a result of a socialization process the characteristics of which evolved slowly over a long period of time in place. In this respect, the moral order to which individuals should adhere is fully transparent to all and is reinforced day after day by the behavior such an order elicits from each member.

Tönnies called this moral order of the countryside that of ***Gemeinschaft*** (usually translated as "community") in order to contrast this traditional culture with the new moral order (or disorder) of newly industrializing cities. Before turning to this, however, it is worth dwelling on this notion of community because it flavors many a view of city life to this very day. It is clear, for example, that Tönnies himself regretted the fact that the moral order of community was giving way to something else. Indeed, from this time forward many others have hankered for a return to this sort of culture, whether by fleeing to suburbia and exurbia or by constructing whole new "**garden cities**" such as Letchworth, England, during the late nineteenth century and, most recently, Celebration, Florida (see Chapter 9). "Community," in other words, continues to be perceived as a more authentically humane, even more ethical, manner in which to conduct human relations which has become quite lost as a result of industrialization and urbanization. To mend the social ills of the latter processes, then, necessitates a return to such community relations, whether by a thinning out of the complexity of industrial and post-industrial city life or by recreating small village life altogether, or, indeed, both at the same time in newly themed citified bits of the modern metropolis.

This same sense of community lost can be discerned in the current discussion concerning the need to reconstruct something called

social capital in the city, as discussed below. Here, however, it is necessary to dwell on this notion of community a bit in order to get a better sense of the full implication of such a hankering for cultural times past. Briefly, a mere glance at the characteristics associated with community very quickly reveals a Janus-face to the very concept. The kind of social relations that such community represents appears to be quite authoritarian and socially conformist in which no deviation from traditionally derived norms can be tolerated. This makes any attempt to do things differently or to act differently than what is considered the norm quite difficult, if not impossible. And so village life continues to replicate the same way of doing things and the same relations of social power over and over again for a social eternity. That this social eternity, in reality, often includes quite hierarchical relations of power, including those based on gender, with quite severe social sanctions against transgressors of the moral order, renders such "communities" perhaps less something to recover and more something to avoid, if possible.

CREATING A NEW CITY-BASED "SOCIETY"

Anyone who has lived in a small town or village surely can relate to this less flattering account of the moral order of *Gemeinschaft*. Indeed, one of the attractions of city life has always been considered by many to be the escape from such traditional, rigidly hierarchical, socially confining, good-[illegible] was certainly missed, for the most part, in Tönnies's own early analysis; but it is also missed in more contemporary ones. Perhaps this is because most, particularly contemporary, commentators have not experienced small town life themselves. In any case, this flip side nature of "community" is generally ignored, especially when such community is contrasted with what Tönnies considered to be the new cultural relations emerging in newly industrializing cities. Here, he argued, social relationships were no longer based on longstanding traditional or extended family relations but, rather, on the basis of more rational relations, such as contracts as well as brief encounters, in order to conduct one's day-to-day business with relative strangers. Such relations were, in turn, largely organized according to impersonal economic criteria such as efficiency and utility, rather than on longstanding social norms and sanctions. As a result, this new city culture, or what Tönnies called ***Gesellschaft*** (usually translated as "society"), was no longer molded by previous social norms and values but, instead, had to evolve its own mold largely on the basis of formally institutionalized, impersonal codes of conduct (laws, legal systems) and forms of social sanction (police, courts).

Remember that in Tönnies's time and place, the rapidity by which industrial urbanization was occurring was quite unprecedented. It did, indeed, seem like there was an entirely new type of human culture emerging in cities. Strangers rapidly thrown together in dense groupings simply had to find a way to live together relatively peaceably and regularly. Close neighboring strangers with probably quite different values and interests and working at new-fangled industrial jobs were largely unknowable in the hustle and bustle of newly mechanical, sped-up, day-and-night clock time of the ever lighted industrial city. Indeed, families themselves were increasingly torn asunder as there was no room in city housing for extended families and there was an increasing separation of workplace from living space. Furthermore, as mechanical work time extended to most of the daytime hours, former household services had to be purchased from strangers at the laundry, the tailor, the barber, the doctor's office, and in the bars and restaurants. One simply had to live, and interact, face-to-face with strangers in order to get by. In fact, paradoxically, one had actually to depend upon strangers for day-to-day needs. Yet, in the end, unknown strangers are essentially difficult if not impossible to trust.

Such a situation calls out for the impersonal forms of social norms and sanctions mentioned, of course. But it is elsewhere where the importance of Tönnies's message has resonated with city commentators to the present day. What he emphasized was that this transition from rural "community" to the "society" of the industrial city was just that: a transition. It represented a crucial period of time in which one set of cultural norms and sanctions was breaking down and a new set was still in the making. Because of this, it was a period of time whereby individuals, particularly those most new to city life, were likely to have a very hard time understanding what was expected of them in terms of proper and, indeed, permissible social behavior. The necessity of regularly interacting with essentially untrustworthy strangers rendered an understanding of the larger cultural rules of the city game much more difficult to make out in this regard.

This loss of known and even knowable cultural bearings in the transition to city life has led many commentators after Tönnies to suggest that city life itself lends itself to moral confusion on the part of city dwellers. Such confusion, in turn, can lead many to withdrawal more and more into themselves, to shy away from any attempt at real social connection with their city neighbors. To the extent that this is the case, one becomes more and more alone in the densely packed crowd largely without care for others, even to the extent of ignoring both cries for help or outright crimes actually witnessed. Indeed, such aloneness in the city crowd may even instigate one to begin to hate the unknowable crowd as the ultimate mistrust. One thus becomes literally anti-social in one's alienation, perhaps even to the point of behaving in a socially deviant manner. Or, indeed, one becomes so alone in the crowd that one begins to hate oneself, to strive for self-destruction, overcome by the stimulation of the crowd that really does not care.

Alternatively, others, even those new to the city, may see this loss of traditional social norms as more a form of individual liberation. And this, again, is the flip side of both the culture of community and society in the city. This transitional period of losing one set of cultural norms and constructing another allows individuals to make a culture as they see fit. In other words, city culture is not about doing things like they have always been done but, rather, about making things up, so to speak, as one goes. This is why cities have always been the source of social innovation, from technological and organizational change to the latest fashion in clothes. On the very basis of being a relative stranger there are relatively few fixed expectations and therefore few limits on what an individual may do. And this frees every individual to create new ways of doing just about anything, including new ways of relating to other similarly free individuals in the city.

NOT CITY CULTURE BUT CITY CULTURES

Nevertheless, most commentators after Tönnies, particularly those in the United States, have tended to emphasize the darker side of city culture as opposed to this more socially liberating one. This simply has fueled the general suburban mindset of most Americans. Yet, it was also largely commentators from the United States who made it clear that city life is experienced and lived quite diversely by different groups in the city. This was certainly quite apparent in rapidly growing industrial cities in the United States which were absorbing wave after wave of immigrants from Europe and Asia, as discussed below. More generally, the idea of *city cultures* derives from two different considerations. First, there is a difference between observations of the cultural life of cities and this life in reality. Many of the more adverse comments about evolving industrial city culture reflect more the prejudices of the commentator rather than the actual goings on within the city. This remains particularly the case today when poorer neighborhoods are singled out for comment. What looks to some to be a culture of social chaos and haphazardness may not, in fact, be that as most poor neighborhoods have developed their own moral orders. Similarly, it is essentially those who benefit most from the authoritarianism of traditional community who decry the hustle and bustle and cultural openness of city life.

Second, it is clear to most anyone who has spent some time in cities that city individuals do not really live in, or socially navigate, the entire city but, rather, live in much smaller parts of it. Most city neighborhoods, for example, are small communities in this respect, with their own culture. The only difference between these communities and those of traditional villages is that such neighborhood culture is relatively freshly made with a much more diverse set of participants. In this respect, and to the extent that they are successfully established, neighborhood cultures in the city can be considered a social innovation in terms of trust-building among strangers.

This is the potential at least. It may be that such urban villages merely replicate the authoritarian community of rural areas, as is the case many times when city neighborhoods are filled up by people from the same village or the same ethnic group or religion. In fact, this is a common enough occurrence which results from what is called **chain migration** to cities, where people from the same rural area or ethnic group come to inhabit the same part of the city for mutual support. Nevertheless, even in this situation the surrounding city environment tends to transform such community relations as members come into regular contact with city strangers in nearby neighborhoods. This is particularly the case over time as more individuals within the community are born in the city, as well as when other non-community members move into the neighborhood.

A CITY CULTURE OF POVERTY

But this notion of neighborhood villages within the city does underscore the ways in which culture in the city is more diverse than most commentators suggest. Most visibly, for example, there are cultures created not just by spatial proximity, but also on the basis of how individuals variously experience city life in terms of their social status. There are, in other words, quite different cultures of poverty and wealth in the city. In the United States, this is the difference generated by living in the inner city as opposed to the suburbs or the gleaming, newly gentrified central cities and CBDs. In Europe and most of the Global South, this is the cultural difference that emerges from living in the poorer suburbs far away from the riches of central cities (see Chapter 9). The point is that the culture of poor areas of the city is one in which individuals have created means of coping with their less fortunate social circumstances in ways that are hard to understand by those more fortunate. Indeed, as alluded to above, for the most part this culture of coping has been understood by the latter as a form of deviance or otherwise anti-social behavior at its core.

Without repeating the history of city poverty discussed in previous chapters, it should be clear that the city poor must contend with very adverse conditions of job loss and a quite significant reduction in social services as a result of contemporary neoliberal policies. In turn, because of the fiscal squeeze brought on by the decline in tax revenue and the necessity of entrepreneurial city policies, the built environments of poor neighborhoods have deteriorated and everything from roads to sanitation services to, most significantly, schools have suffered from lack of adequate funding. As a result, the poor areas of most cities have generally become caught in a self-propelling spiral of decline with fewer opportunities for jobs and for the acquisition of the skills necessary to fill those jobs that do exist.

At the same time, access to other parts of the city has become increasingly difficult for the city poor as more and more of the central city has become gentrified and the gentry has demanded more and more security. This has led to a very significant decline in open public space as city shopping takes place increasingly in closed, privately policed, boutique malls, offices are conjoined by skywalks, and adjoining streets, gardens, and parks have come under increasing security camera surveillance. In turn, because of the necessity of catering to the needs of the gentry, what public space continues to exist is policed much more vigilantly to make sure that it is free of the homeless and beggars and drug dealers and others who do not fit the image that cities need to project in order to attract this now global gentry. Particularly in the United States, most city poor also have little access to richer suburbs as public transportation is either inadequate or non-existent, leading some to talk of the actual **transit captivity** of the inner-city poor.

In this context, the culture of the inner city is one of shared adaptation to such lack of opportunity and even mobility. This is experience of the city as a mostly hostile place with little room for escape to a better life situation. It is no wonder that poor city neighborhoods continue to be sources of periodic and quite active unrest in cities, as witnessed most recently in London and other major cities in England. Indeed, that the spark for such unrest most always has come from what local people perceive to be overzealous policing of their neighborhoods does not bode well for the future. Such policing arguably has increased in extent and intensity as city after city now seeks to portray itself as a safe and fully pacified social environment for the global gentry in which to live and work.

Finally, one can argue that even some of the details of inner-city culture make sense, from street gangs for mutual protection of property and person to dealing drugs and other illicit contraband as an informal, yet necessary, means of livelihood. Indeed, the latter activity exhibits well the kind of entrepreneurial activity that, given the circumstances, should be rewarded rather than sanctioned by city authorities as it would be in the formal city economy. More generally, some have argued that the city poor have constructed a **culture of poverty** the behavior and values of which simply reinforce and reproduce the very poverty from which this culture arises. While there is some truth to this, this conception has been mostly used as a rather thin veil for blaming the victims for their own misfortune instead of focusing on the greater social, economic, and even built context that elicits a certain kind of social behavior as a cultural means of adaptation.

AN INCREASINGLY PRIVATE AND GATED CITY CULTURE OF WEALTH

On the other side of the cultural coin, since the beginning of industrialization, more wealthy city people have tried to escape its results either by migrating to non-industrializing suburban areas or, as in the case of older cities of Europe and most post-colonial cities of the Global South, by prohibiting industry from taking over central areas of the city. In this way, those with the means to do so successfully avoided as much as possible the rapid onslaught of new city people of much more limited means and their housing, as well as the congestion and pollution of newly built industrial factories. The newly built or preserved neighborhoods of the rich were increasingly and actively segregated from the rest of industrializing cities and such segregation was often enforced by city policies and the actions of housing developers and gatekeepers. Spatially distinct city cultures of wealth thus emerged early in the industrialization period and, of course, continue to exist in today's metropolitan areas.

Like city cultures of poverty, city cultures of wealth are created by individuals facing similar circumstances. Unlike city cultures of poverty, however, city cultures of wealth are created by individuals with much more control over these circumstances. The rich experience the city largely on their own terms with regard to opportunities for employment and residence. Their ability to access all parts of the city is also much greater both in terms of the ability to procure transportation and to avoid limitation of entry due to social status. In short, the spatial segregation of the rich in the city is obviously a conscious self-segregation in an attempt to maintain their wealth and, then, status in now post-industrial cities.

As described in previous chapters, this self-segregation of the city rich has been enforced, from the beginning, by actively discriminatory housing policies, whether based on income or ethnicity or, indeed, both. This continues to mean that the attempt to escape the more diverse, densely packed population of the city has resulted in the creation of more simplified communities of people more or less alike, with similar material circumstances as well as social beliefs. In this respect, the culture created by the city rich is somewhat like the *Gemeinschaft* conceived by Tönnies and, indeed, many developers and promoters of places for the city rich today suggest as much.

It is worth dwelling on this last point. From early streetcar suburbs to today's master-planned, full-scale central city, suburban, and exurban "communities," the social relations of the city rich have, indeed, exhibited *Gemeinschaft* characteristics. Yet, as noted above, when stripped of the largely romantic nostalgic connotations given them by Tönnies and followers, such characteristics are not necessarily good in terms of human social relations. The authoritarianism and conformism and resistance to change or social innovation such community normally entails should give one pause here, at least. But, then again, this might be the material point. If the main post-Renaissance goal of the city elite has been to maintain social status in the midst of the many economic changes and resulting transformations of cities, then securing and increasing their riches has had to be their number one goal. The safety of their persons and, as importantly, their property must be seen, then, as the top priority in their locational decisions within the city. From this perspective, the best way to do this is to seclude themselves from the larger city population and enforce this seclusion in the most forceful way possible. If this means that only the relatively rich can join in this seclusion, so be it. If it also means strictly enforcing conformity in housing styles and property use and even social behavior in these communities in an authoritarian way, so be it as well. This is the social price to be paid for securing one's wealth and then status in the city.

But this sort of *Gemeinschaft* is even thinner than that enunciated by Tönnies. For this author, such community at least involved all the people of the village, no matter how rich or differently situated in village life. The communities of the self-segregated rich today are not even as socially diverse as this. Moreover, such self-segregated, purposely thin city communities have become quite self-replicating, particularly in these times of neoliberal, entrepreneurial city development. The key is the quest to have whatever tax dollars they owe be available for their own needs and not those of others in the city. This quest has resulted in special business improvement tax districts in central cities and CBDs, as well as the political incorporation, or outright privatization, of large swathes of suburban and exurban territory in order to avoid political annexation to traditional cities. And so the suburban and exurban communities of the more fortunate enjoy the best services and, importantly, provide the best educational opportunities that their money can buy as traditional cities increasingly suffer fiscal squeezes.

Among these self-segregated city people, private housing and commercial development corporations now dictate "community" rules and regulations and homeowners' associations collect fees to enforce such rules, all with the goal of upholding local property values above everything else. In the United States, this most recent trend in self-seclusion has been called the rise of a **privatopia** for more fortunate city people as they no longer need to think or care about anyone else's needs but their own. As a result, citizens of such privatopian metropolitan worlds no longer even really know much about the plight of less fortunate city people or, really, any others outside of their gated communities. Such citizens now not only live but also work, get educated, play, and otherwise spend their leisure time almost completely among their own kind in privately financed, socially restrictive facilities. Indeed, as described in Chapter 9, the very latest trend in the self-seclusion of more fortunate city people is the development of vast master-planned, full-scale townscapes including different varieties of housing, business districts, medical facilities, recreational areas, and so on, either fully gated or otherwise physically and socially set off from the surrounding metropolitan area.

A CITY CULTURE OF ETHNICITY

The onset of the continual process of rural-to-urban migration as a result of industrialization also brought with it the prospect, and largely the reality, of rural immigrants of much ethnic diversity. Cities, in this respect, became much more multicultural than previously and this brought with it the necessity of finding ways to negotiate sometimes quite sharp cultural differences now rubbing so closely into each other. Such cultural differences were perhaps most noticeable in rapidly growing U.S. cities as waves and waves of European immigrants flowed into them in the latter part of the nineteenth and early twentieth centuries. Unlike the earliest European immigrants, these newer immigrants, particularly Central European Jews, Italians, Sicilians, and Irish, did not flow right through cities toward the western agricultural frontier but, instead, came to settle more or less permanently in major cities of the

Northeastern Seaboard. Similarly, on the Pacific Coast more and more Asian immigrants were settling, particularly in the growing cities of California. Industrializing cities in the United States thus became multicultural extremely rapidly as whole districts took on immigrant village-like status as ethnic groups huddled together for self-protection and -promotion in their new environs. Indeed, that the vast majority of these newer immigrants came with very little money to their name meant that whatever support that could be mustered from earlier immigrants from their homelands was absolutely necessary for ultimate survival.

Yet, for this very reason, as well as the fact that the newer immigrants did not break for the frontier at the first opportunity, these immigrants came very quickly under suspicion on the part of earlier European arrivals as not quite fit for assimilation into American culture. This story is long and involved and open to much debate. What is most important here, however, is that the rise of little Italies and Sicilies, Chinatowns, as well as Jewish and Irish, etc., ethnic districts in major U.S. cities helped to solidify in the minds of earlier settlers that the social problems of such cities were really the result of the inability or unwillingness of these newer immigrants to assimilate into the wider American culture. Social problems of the city thus became equated with ethnicity early on in the industrial era in the United States, hastening the departure from the city of earlier immigrants and a growing antipathy toward increasingly "colored" city culture, in general.

It is at this point that the safety of chain migration to particular districts of the city on the part of the most recent immigrants became reinforced by outright discrimination and **ghettoization** by city authorities. Newer immigrants increasingly were forced to stay within their districts by discriminatory housing and work restrictions, if not simple fear of violent reprisal. In this respect, the lack of assimilation into the greater American culture became a self-fulfilling prophecy of outright lack of opportunity to take part. This became even more apparent when the descendants of African slaves began to migrate northward toward industrializing cities in search of economic opportunity. No longer slaves and lacking opportunity as (formally) free people in the still rural South as a result of overt racism, African-Americans came in droves to industrializing Northeastern and Midwestern cities in the early part of the twentieth century. Known as the "Great Migration" from the South, African-Americans filled up central city locations close to new mass-producing factories in cities such as Cincinnati, Cleveland, Pittsburgh, Detroit, and Chicago. They also arrived *en masse* in Baltimore, New York, and Philadelphia. This was a major cultural change in these cities, adding a color that was always a part of American culture but never allowed to assimilate fully with it. In fact, African-Americans were actively disallowed from culturally assimilating; yet, here they now were literally rubbing shoulders with other city people in the factories of rapidly industrializing cities.

This rapid urbanization of African-Americans simply reinforced the growing perception of industrializing cities in the United States as caldrons of potential and actual social conflict and other trouble. Previous discrimination against newer immigrants from Europe now blossomed into outright racism against people who, while now free, still exhibited the legacy of sub-human slavery in the color of their skin. Even more so than Europeans, African-Americans were discriminated against on all fronts and, as a result, were literally ghettoized in specific districts of the inner city. Indeed, such reinforced discrimination against African-Americans in the city actually loosened that against most European immigrants (but not Asians and, eventually, Latinos) as these were allowed greater opportunity as now "better" candidates for cultural assimilation.

This loosened discrimination allowed later generations of European immigrants to blend into the greater metropolitan area. Over time, even most Asian immigrants have done the same. While little Italies and Chinatowns, etc., still exist in major U.S. cities, these are much less a necessity of survival in the city than a celebratory façade for consumers of city culture. Unfortunately, this is not the case with regard to the African-American-dominated inner-city areas of most every major U.S. city. While formally outlawed, discriminatory housing policies still operate and African-Americans still find their ability to negotiate the entire city of opportunity extremely circumscribed. While those African-Americans of more financial means have been able to escape the inner city, there still exists considerable segregation of housing on the basis of color in the suburbs and exurbs. For those less fortunate, the inner city remains the only option and, with the recently rapid decline of city-based industrial jobs, an increasingly dismal one at that.

MULTICULTURAL CITIES IN EUROPE OR THE EMPIRE STRIKES BACK?

Because of their longer history *vis-à-vis* that of the United States, most European countries have much more ethnically homogeneous populations. Industrial urbanization thus was less an ethnically diverse affair and more one of the shock of city life as opposed to that of the country, as discussed above. Yet, in the latter part of the twentieth century, this began to change. Many more immigrants from imperial, and former imperial, possessions began coming to Europe following the same pattern of chain migration and city settlement as those who came to the United States. Immigrants from the Americas, Africa, and all parts of Asia began filtering into European cities, sometimes in great numbers. In fact, many countries, particularly in the more industrialized northern and western regions of Europe, actively recruited such immigrants because of growing industrial labor shortages as a result of the war effort. These immigrants were nothing if not culturally diverse and many times shockingly so to most Europeans. Vastly different cultural behaviors and mores now became quite noticeable as they rubbed together closely in historic European cities.

As major districts of European cities filled up with former colonial subjects of much cultural and literal color, the perception of cities on the part of many Europeans began to mimic that of most Americans. Suburbs and exurbs thus became increasingly attractive places in which to live. More importantly, these recent waves of immigration to European cities have rendered issues of ethnic relations and cultural assimilation much more important than previously. And here is where the experience of the United States is pertinent. To what extent, for example, are Europeans allowing the opportunity for these new immigrants to assimilate into the receiving culture? Very recent riots in major cities of England and in the industrial suburbs of cities in France suggest that there has not been enough effort in this regard. Indeed, the apparent spread of militant anti-Western Islam among disaffected immigrants and their progeny in Europe further suggests this to be the case.

Yet, it is too early to tell how things will turn out in European cities. The culture shock of so much influx of cultural difference from former imperialized regions so rapidly is still playing itself out.

It just may be that better opportunity for housing and jobs and political representation and therefore better cultural assimilation for recent immigrants will be forthcoming much like it eventually was for Italians, Jews, and Irish in U.S. cities. However, this is not a given. It just may be, instead, that opportunities such as these will not arise or arise inadequately in terms of actual need. Especially in these neoliberal times, many more Europeans have come to see such immigrants as less a valued labor resource than competition for ever-scarcer jobs for Europeans, as well as a growing drain on already dwindling welfare state services. If this is the case, then the new immigrants to European cities likely will be faced with the same kind of continuing discrimination and spatial segregation as that of African-Americans and, most recently, many Latinos in U.S. cities. Indeed, the longer experience of Turkish immigrants to German cities does not bode well in this regard. If this, in fact, is the case more generally, then the recent riots in European cities noted above may just be the tip of a much larger, still mostly hidden, iceberg of potential urban unrest.

THE POST-INDUSTRIAL PACKAGING OF PLACE

Currently, culture, in general, is considered an increasingly important factor in terms of the overall development prospects of cities. Competition among cities to attract post-industrial investment and people of the right tax bracket has rendered it more and more necessary to package cities in the right way. This includes whatever cultural difference that might render them unique and potentially desirable places in which to live. However, since all cities must now be entrepreneurial in this way, they already more or less offer the same sorts of generic amenities to attract the global gentry and their money, as discussed in previous chapters. Because of this, the growing belief is that something unique about a city's culture or history must be found and promoted in order to set it off from its competitors. Whether this is a city's former role as a working seaport, a site of important political events, a capital of ethnic celebration, or even a site of multicultural relations among little Italies and Chinatowns, etc., such traditional culture must be recast as uniquely significant in today's competitive marketplace.

But this celebration and promotion of city culture must be understood for what it is – that is, primarily a means to attract the global gentry. As such, traditional city cultures must be rehabilitated in ways that are not too authentically different or, indeed, potentially controversial but, rather, in ways based more on the selective amnesia of nostalgic memory. Put differently, only cultural differences that are not too different and therefore ultimately safe for all to experience as entertainment without much critical thought are the kinds of cultural difference most promoted. In this respect, the packaging of place culture is more akin to what Disney calls "imagineering" than it is to any sort of authentic rendering of traditional city culture. Imagination plays as big a role in this regard as historical fact. And just like Disney, in order not to spoil the intended experience of this imagineered cultural environment, strict restrictions must be placed on who has access to such an experience as well as on how those who do actually, in fact, experience it. This is merely another aspect of the ongoing privatization of cultural life in the post-industrial city.

Box 7 Gender and the city

Industrialization brought forth much social change. Cities began to grow continually and rapidly and city life was substantially different from agrarian life. Gender roles were also transforming. On the farm, men and women had separate but largely equal roles in ensuring the reproduction of households. This meant that women's work was considered equally important. Of course, **patriarchy** ensured that the maintenance of the home and child rearing were considered a woman's natural nurturing role; but farm chores were split on the basis of necessity, not importance.

Industrial city life, however, entailed the emergence of quite different gender relations as a result of two main processes: first, industrialization resulted in the explicit separation of productive space from living space and, then, formal production time from household reproduction time. Living space and time thereby became privatized and relatively de-valorized, while working space and time were rendered public and relatively valorized as sites and times of "really important" work. These different spaces and times then came to be identified with gender roles. And here is where the complexity of gender and the city becomes most apparent. In the early years of industrial urban development, women (and children) were mostly employed as wage earners in the new factories instead of adult males. Conventional histories maintain that this was because women (and children) were better suited to the newfangled technology of industrial production. In fact, this is still said about women in the growing number of mass-producing manufacturing complexes in cities of the Global South. However, the reality is that women (and particularly children) are more docile as a result of patriarchy and therefore come cheap and with relatively no complaint.

Second, then, these new gendered spaces and times, both in the early years of the first industrial cities and in today's rapidly industrializing city South, made, and make, women's work that much more publicly important than men's work. Indeed, for many women, new industrial work was, and is now, a liberating proposition, allowing a more public role and putting family-independent money in their pockets. And such an empowering, even liberating, role of public work and city life for women has been felt regularly in the history of cities from the very early years of industrialization to current processes. This was most noticeable, for example, during the war years when women found greater numbers of jobs in city factories as menfolk went off to fight. The well-known legend of Rosie the Riveter, in fact, is as much about women's capabilities and independent strength as it is about patriotic heroism.

So, new city work has been liberating for women and this liberation has only been reinforced by bringing many women together in the relatively small spaces of cities where they can socialize around common complaints about, among other things, traditional patriarchal roles. But this is an idealized picture, in many respects. Actual industrial workplaces have been largely hostile to women overall, with long, grueling hours of menial, repetitive tasks, overbearing male managers, and the lack of job security, rendering continued gendered docility a must. Similarly, many studies have shown that married working women still do much of the household management and child care duties, thereby working overtime but not being publicly recognized, or rewarded, for such. For these and other reasons it is clear that while significant spaces and times for women's liberation from patriarchy may be found in cities, this is not a guarantee. Rather, any such liberation must be actively struggled for if it is to come about at all.

FURTHER READING

The topic of gated communities is central to Rowland Atkinson and Sarah Blandy's recently edited book *Gated Communities: International Perspectives* (New York, NY: Routledge, 2006), which, significantly, explores this issue on a global scale. Sylvia Jenkins Cook examines important aspects of gender liberation in the city in her *Working Women, Literary Ladies: The Industrial Revolution and Female Aspiration* (New York, NY: Oxford University Press, 2008). Other recent books on gender and the city include Judith N. DeSena's edited *Gender in an Urban World* (Bingley, UK: Emerald Group Publishing, 2008) and an earlier edited volume by Judith A. Garber and Robyn S. Turner, *Gender in Urban Research* (Thousand Oaks, CA: Sage Publishers, 1995). The increasing importance of packaging city places to attract increasingly transient people and capital is explored variously in Gerry Kearns and Chris Philo's edited volume *Selling Places: The City as Cultural Capital, Past and Present* (Oxford, UK: Pergamon, 1993). Evan McKenzie explores the relatively rapid growth of privatized spaces in U.S. cities in *Privatopia: Homeowners Associations and the Rise of Residential Private Government* (New Haven, CT: Yale University Press, 1994). Classic articles on city cultural life can be found in Richard Sennett's edited *Classic Essays in the Culture of Cities* (New York, NY: Appleton-Century-Crofts, 1969); and, finally, David Wilson does a fine job of depicting racial relations in contemporary U.S. cities in his *Cities and Race: America's New Black Ghettos* (London: Routledge, 2007).

8

CITY ENVIRONMENT

The emergence of cities represented a major transformation in the way in which humans relate to non-human nature. Already, the development of agriculture marked growing human control over natural processes. Hunter and gatherer existence meant that humans more or less lived off the bounty of nature as it presented itself. Agrarian life, on the other hand, signified humans taking charge of their own material reproduction by taking better charge of nature itself. The rise of cities signified even greater success in this, as discussed throughout this book. Now actual surpluses of human-controlled agricultural production were available which allowed ever-greater numbers of people to live off nature without actually controlling it themselves. Hence, human civilization was born.

This book has been devoted to details of this transformation, to be sure; but now note four specific things of importance to the city environment. First, the ability to control nature for human purposes already transforms natural processes. As agriculture spreads and intensifies, this transformation becomes even more extensive and profound. Biodiversity, for example, diminishes as certain grasses such as wheat and barley are singled out for cultivation and other plant species become "weeds" to be eradicated. Natural waterways are modified, often dramatically so, in order to supply cropland with irrigation. Forests are cut to provide more and more land for cropping. Some animals are domesticated, which both disrupts natural food chains and renders such animals wholly dependent upon humans for survival and even evolutionary characteristics. In short, the development of agriculture signified a stunning emergence of human control and therefore modification of non-human nature. As this evolved into the capacity to support even non-producers, this modification of nature was greatly intensified and spread ever more widely.

Second, the development of cities and the emergence of city people brought with it new kinds of modifications of non-human nature. Land had to be cleared and natural terrain transformed in order to make possible humanly engineered structures for day-to-day human social activities and shelter. Such structures were built in close physical proximity as large groups of people came to reside on small pieces of territory. This marked a major transformation of natural landscapes and the new densities of human populations put extra demand on surrounding nature to supply their various needs, from water and food to the absorption of human waste. Cities, in other words, became specific sites of intensified human demand on non-human nature merely because of the new and growing crowds of people they entailed.

Third, city people are also differently related to nature as a result of their very existence. City people are essentially consumers of the products of nature that are controlled elsewhere. It is in this way that cities are essentially parasitic on their rural hinterlands, as noted previously. This means that, for cities to grow in population, more and more natural hinterland must be taken under human control and therefore transformed. In today's globalizing world, this relation of cities to nature has come to be understood as the variable **ecological footprint** they each make on the planet.

Finally, it should be noted that city people, as a result of the very way in which they live day to day, necessarily come to understand or, indeed, increasingly *mis*understand the human relationship with non-human nature. City people are literally surrounded by human productions, whether in terms of the built environment or the social environment consisting of all the cultural emergences of human civilization. Arguably, this generally renders city people quite alienated from, or at least increasingly ignorant of, the very natural processes that support their continuing existence in cities. In fact, some argue today that because of this effect of city life, in general, the future of global nature does not look very promising at all in a world that now consists of a majority of city people more and more ignorant of natural processes.

EATING, DRINKING, AND MAKING WASTE IN THE CITY

Now is the time, however, to begin to determine more precisely the overall environmental effects of cities. As mentioned, for cities to exist at all, city people must find ways to feed themselves. This entails the extension of agriculture, as mentioned, but also the development of more effective modes and networks of transportation to get surplus agricultural products to the city. As cities grow, more and more land must be brought into cultivation to meet this need, which, in turn, puts pressure on existing transportation modes and networks to facilitate this process. In this way, growing city demand for food establishes a self-propelling process which modifies greater and greater swathes of natural area for human needs.

City people also need freshwater, both for their very survival and also for the successful cultivation of agricultural crops. As a result, many early cities arose on river and lake banks with already fertile soils and plenty of water. But as city populations grew, these sources of water were increasingly diverted from their naturally occurring paths toward city and irrigation uses via canals and other human constructions such as aqueducts and pipes. Similarly, groundwater sources of water were increasingly accessed by ever-deeper and larger wells. This growing modification and use of freshwater sources necessarily disrupted natural processes and relations on an increasingly grand scale as the number of cities and city people continued to grow.

The environmental effects of cities can thus be measured in terms of the extent of food and water demand and how, and from where, this demand is met. But this is only one part of the story. The consumption of food by densely packed city populations results in concentrated sources of waste products, from leftover or unused food products to, of course, human waste. Space must be found, and processes must be created, for somehow disposing of such waste which otherwise would eventually bury the city in which it accumulates. While this seems obvious, the key is that cities produce such vast concentrations of such waste that wherever it may be "dumped" it quite overwhelms naturally occurring systems at such sites. And this is if the collection and dump of such material is successful. If this latter is not even a sure thing, as in many historical cities as well as in poorer post-colonial cities today, then such waste accumulation is spread throughout the city, breeding all sorts of disease-carrying microbes and attracting many nefarious, equally disease-carrying, insects and vermin.

Disposing of human waste in such concentrations has been a particularly difficult problem for cities throughout history. The development of an adequate sewer system for entire city populations has been only a very recent historical evolution, as discussed below, and remains quite unevenly spread among cities even today. Flushing human waste out of the city necessitates large and continuous supplies of water properly diverted in order to be effective. Good engineering and construction of conduits is also necessary to make sure that adequate mechanisms for the collection, transport, and ultimate disposal of human waste are created and maintained. Otherwise, city people would be literally drowning in their own accumulated filth, which attracts and carries its own pathogenic microbes and easily fouls city water sources, from rivers and lakes to underground wells.

Of importance is that the creation of waste-disposal systems within cities is a time-consuming and ultimately expensive proposition. It will certainly not be undertaken by elite individuals within the city unless they are ultimately forced into such a task. Such individuals are more likely to take care of their own needs at best rather than those of the wider city population. And so early sewer systems usually were quite selective, covering only small parts of the city with the rest of the population somehow taking care of their own needs as best as they could. But this is where the effects of city life itself come dramatically into play. Once city populations grew larger than merely the small numbers of elite in any given society, all the population of cities became vulnerable to what happened in any one part of such cities. Disease spread very rapidly and quite indiscriminately through greater city populations either via a contaminated water supply or accumulated waste or, of course, both at once. This was particularly the case in the largest, most densely packed cities of the great empires. Because of this, it soon became clear to city elites that personally financed protection against such environmental threats was not enough. Only more public works for as much of the city population as possible would offer the fullest protection in this regard.

At this point, then, it is important to highlight two things. First, whatever public works for water supply and waste disposal came into existence in cities were the result of social pressure on city elites to act on behalf of the greater city population. Obviously, then, this action did not come entirely voluntarily on their part and, for this reason, most early public infrastructural works remained inadequate and quite unevenly distributed throughout the city. Indeed, this remains the case today in many post-colonial cities. Second, such public works cost money to build and maintain. Again this is obvious and obviously related to the first point. But here it is useful to dwell on this because it was really the absolute necessity of building such works that rendered it necessary, in turn, eventually to establish some other source of public financing besides simply the relatively small numbers of city elite. Tax systems had to be established and maintained for cities to function, which resulted, in turn, in city tax payers with a new stake in how their tax money actually was spent.

Both of these emergences signify an increasing pressure from other city people on the decisions made by city elite. And this pressure has only grown throughout the history of cities. But there is one last thing to underscore with regard to such infrastructural public works in the city. Because they are expensive to build and maintain it has been a struggle to ensure the adequacy of such works for entire city populations. This was true for historical cities, to be sure; but it remains the case today in both de-industrialized cities of the Global North and rapidly growing megacities of the Global South. With severely dwindling or completely non-existent city tax bases from which to afford to create and maintain such systems, some other source of such funds is necessary to avoid extreme public health crises of growing numbers of city people today and in the near future. Given today's neo-liberal, globalized context in which such cities are now evolving, finding such an alternative source of public finance has become highly problematic.

THE URBANIZATION OF NATURE ON A GLOBAL SCALE

So cities can first be thought of as needing environmental inputs and producing outputs which, in turn, impact upon the natural environment. As empires and imperial cities grew, nature was thus affected to an ever-greater spatial extent. But this growth of city-derived environmental impact was not continual nor, of course, planetary in scope until much after Western overseas expansion commencing in 1492. Before that, cities did not grow continuously, but rather in fits and starts in specific areas of the world. Cities grew with empires as additional land was conquered and brought into cultivation; but they also declined as empires fell. So, for example, the environmental effects of agriculture mentioned above only really affected small areas of the world in relative isolation. After 1492, however, Europeans began to create a world system of production and consumption resulting in vast environmental impacts all over the planet.

Very soon after 1492, for example, the nature of what came to be known to Europeans as the Americas was intensely transformed. Vast mining operations were constructed in Mexico and Peru to extract tons and tons of gold and silver for use in Europe and for European trade with Asia. Domesticated horses, pigs, goats, and cattle were introduced in great number by the Spanish and Portuguese, thereby severely disrupting American ecosystems. Plantation-style agriculture was developed on a massive scale, producing sugar, tea, tobacco, and cotton on the most fertile land in America for export to Europe. In Africa, the growing European-directed slave trade disrupted traditional trade routes and production patterns and, in the South, a new agricultural production pattern emerged more suited to European tastes and trade interests. In Asia, most early contact by Europeans was coastal, although some areas very soon came under direct European control and were put to cash crop production (particularly tea and eventually cotton). As noted in Chapter 3, however, merely the growing European demand for agricultural products and spices modified indigenous production patterns considerably, particularly in agriculture.

This growing city-led ecological footprint of Europe across the globe intensified considerably after the Industrial Revolution (ca. 1750).

Now non-European nature was modified on an enormous scale to produce not only more and more consumer goods, such as sugar, tea, and tobacco for growing city populations in Europe, but also industrial raw materials, particularly cotton for textiles and groundnut and palm oils intended for the lubrication of industrial machinery. This growing urbanization of nature on a global scale included huge swathes of territory in the Americas, Africa, and Asia, now tied forcibly and directly to the industrial development needs of Europe. This greatly accelerated the industrial development and resulting population growth of European cities which were already well on the path to industrialization via the intensification of demand for such inputs from their more immediate rural hinterlands.

Industrialization also brought with it an ever-increasing amount of city output that affected the environment. The growth of city populations due to continual rural-to-urban migration vastly increased the amount of city waste that was produced. At the same time, new industrial processes were creating new types of waste products, especially chemical effluents from bleaching and dying textiles and animal skins, as well as from the lubrication of industrial machinery. Flushing such waste from the city proved an even more arduous task than that of human waste as such mostly toxic chemicals did not biodegrade very easily or, then, rapidly. Similarly, as industrial cities were fueled increasingly by coal and then petroleum, air pollution became a serious problem, particularly as smoke from ever-increasing numbers of house and factory chimneys mixed with fog to create city-smothering **smog.** Yet, during this early period of industrialization, such environmentally adverse outputs from cities largely affected the city itself and its immediate rural hinterland. In today's world of planetary urbanization, however, such nefarious environmental outputs now have global reach as part of the much larger ecological footprint of cities overall.

THE CITY AS AN ENVIRONMENT

This broadly historical overview of ecologically important inputs to, and outputs from, the city helps to frame the study of the environmental impact of cities in general. Now it is time to consider more specifically some contemporary environmental issues deriving from ever-expanding city life. The very creation of humanly built structures, for example, significantly modifies natural processes, as noted above. Built structures in such close proximity within cities simply magnify this effect. The constructed surfaces of houses, factories, office buildings, and densely constructed city roads, for example, create an interlinked mass of impermeable surface which alters the environment considerably. Such a surface does not allow for the absorption of precipitation or for the *in-situ* degradation and absorption of human and other organic waste. The combined effects of this are that, after precipitation events, increasingly polluted city water runoff flows with increased volume and rapidity across such surfaces and collects in whatever breaks in impermeable surface exist or, indeed, in intra-urban waterways. Such flows and collections of waste-filled water, in turn, become breeding grounds for disease vectors, including insects, and also quite overwhelm the ecosystems which exist in intra-urban water bodies.

The built environment of cities also alters the circulatory and speed patterns of wind, which become entangled and adapted to street patterns and the various heights and angles of humanly constructed forms. This alteration of wind patterns helps to create what is called an **urban microclimate**, which can differ quite substantially from surrounding naturally occurring climatic conditions. At the same time, the specific way in which wind pattern is altered via obstruction in city environments generally renders wind much less useful in renewing city air by flushing out airborne contaminants and other pollutants.

Finally, the densely packed built environment of cities changes the rate at which sunlight and heat is either absorbed in the surface or reflected. Generally, the more city surface that is humanly constructed, the more heat is absorbed. This is the most significant source of urban microclimatic difference. Because of higher heat absorption, or what is known as lower **albedo** or reflection percentages, cities tend to be warmer than surrounding rural hinterlands. Everything from built surface material (bricks have higher albedo than concrete and asphalt), surface color (white and other light colors have higher albedo than black and other dark colors), and surface texture (smooth surfaces – except windows – have higher albedo than rough or uneven surfaces) affects the absorption rate of solar energy. Since most cities continue to be built with little consideration of such environmental effects, much attention recently has been put on what has been called the **heat island** effect of cities or districts within the city. Such heat island effects, particularly as these are exacerbated by carbon-based emissions from industry, automobiles, and other city-based energy usage, have created in most contemporary cities a never-ending cycle of increased energy use for cooling purposes, as well as localized air pollution. Such pollution not only affects the health of city people, but more often than not spills out over the surrounding hinterland and, especially in the case of the largest sprawling cities, over even greater distances regardless of formal political borders.

PROVIDING WATER AND SEWERS FOR THE ENTIRE CITY

As mentioned, supplying city people with adequate clean water for drinking, bathing, and flushing waste has been a problem from the very beginning of city life. As natural waterways were diverted or otherwise tapped for this purpose, intra-urban water flow and quality increasingly was controlled by human constructions and distribution policy. Again, in historical cities, water was first and best supplied to the city elite, with most other city people left to procure their own water. This, unfortunately, remains largely the case in post-colonial cities today, as discussed below. Yet, historically, this soon became a problem because of the very nature of city life. Very soon, for example, intra-urban waterways, both natural and constructed, became quite fouled with the waste generated by densely packed human populations. Human and animal waste, along with growing manufacturing and construction waste, accumulated and eventually scooted across the new impermeable surfaces of city streets, either by sweeping, collection, or as flow after precipitation. The common belief, if thought was given at all, was that natural waterways and basins were natural filters for such waste and therefore could act as continuous sinks for it.

Intra-urban sources of freshwater from water wells to rivers, lakes, and bays, however, rapidly became festering sewers of filth. As this water became increasingly contaminated in this way, even city elites became susceptible to major waterborne diseases such as plague, cholera, and typhoid. At the very least, all city people came to suffer from such things as gastroenteritis, dysentery, and extended diarrhea. Over time, then, and once the source of such diseases was better understood, city elites had to find ways to provide better water supply and sanitation services for the entire city population, if only for their own survival, even as they moved increasingly to the suburbs. Furthermore, with the rise of merchant and then industrial cities in the West, the characteristics of city populations changed so that middle and lower classes came to participate more in, and eventually dominate, the governance of cities. As this happened, organized resistance to elite domination arose and majority demands for better city services increased dramatically.

As a result, over time most cities in the Global North created more adequate systems of water supply and sewage control. It is important to underscore, however, that this did not come easily, nor was it evenly spread among all cities. It involved much political agitation within the city on the part of non-elite city people, which created enough of a threat from below to get elites to act. And it also should be noted that such threats, along with the fact that adequate water and sewer systems cost a lot of money to construct and maintain. Adding all of this up, as well as keeping in mind the continuing fiscal problems of most post-industrial cities today, it is no wonder that cities continue to have problems with water pollution even today, and even in the Global North. Ever-new chemicals used in the production of goods, ever-more sprawl of impermeable surfaces, ever-more consumption for drinking, bathing, watering lawns, swimming pools, and so on, continue to foul local intra-urban waterways.

But the most significant problem today in terms of city water and sewage is found in the rapidly growing cities of the Global South. Indeed, this is a social and environmental problem of global proportions. The water and sewer systems of the over-urbanizing megacities of the South were most recently constructed by European imperial powers to supply first either Europeans or their native allies put in charge of imperial possessions. These were the city elite, largely segregated within such imperial cities in special districts away from the native population. Such systems, then, were always inadequate for the entire city population and once city populations began to grow quite rapidly after World War II, such inadequacy became more and more acute. Intra-urban waterways and basins are in incredibly bad shape in the post-colonial Global

South today and getting ever worse as increasing numbers of people in need of water and outlets for their waste continue to fill up and expand megacity shantytowns and slums.

Moreover, because of the colonial legacy, the overwhelming majority of the populations of today's post-colonial megacities have no real say in city governance, nor do most of them even work in the formal economy. They are, in this respect, quite off-the-books, or otherwise unaccounted for, in the formal politics and economy of the city. As a result, city authorities tend not to consider themselves overly accountable to them. Most city elites in such megacities thus have largely assumed the role of former imperial masters in this regard. What water and sewer infrastructure does exist, exists mostly for those relatively few who can afford it. The rest of the city population, just like in the earliest cities, must then fend for themselves.

This is obviously a situation of dire environmental consequences as ever-growing billions of megacity slum dwellers foul the very water that they drink and bath in, not by choice but in order merely to survive. The fact that tax- and time-consuming environmental regulation of economic activity, particularly in the informal sector, is extremely lax in the Global South means that not only human but also all kinds of other waste is dumped into intra-urban waterways and basins on an increasingly enormous scale. Finally, this dire situation is rendered even more acute by the fact that the global spread of public health and medical knowledge and technology after World War II has made it possible for city elites in the Global South to protect themselves medically from contagious waterborne diseases like those listed above. In other words, even this disease threat from the city population "below" is no longer an instigator of action on their behalf for the entire city population. And this, surely, does not bode well for the rapidly growing numbers of non-elite city people in the already over-urbanizing Global South.

TRASHING THE CITY

Another environmental issue of importance for cities concerns what to do with all the material garbage, or solid waste, that is generated in the very process of living. This becomes a problem very quickly for cities, of course, with dense populations on small territories. Such waste has to be collected and distributed elsewhere or, again, the city would be very soon covered in it. Throughout history, the necessity of finding somewhere to deposit such city waste has led to ever-expanding boundaries of cities. This suggests already a continuing environmental problem for city dwellers. As cities expand in territory due to population growth, much of this new territory is likely to have been a former dumping ground for garbage, with all that this means in terms of lingering disease-bearing microbes and vermin for new residents. Unless this former dump has been properly cleaned up or sanitarily covered in some way, in other words, such people are likely to suffer disproportionally from gastrointestinal parasites, worms, and diseases such as plague and yellow fever. Again, these are public health issues important for entire city populations and eventually came to be understood as such throughout the Global North as appropriate actions were taken to more suitably rehabilitate such sites. As noted below, however, such health issues remain a problem in cities in the Global South.

Over time, there has evolved three main ways in which to dispose of the solid waste of cities, each of which has specific environmental impacts. The first is most obvious – that is, open-pit dumping in which some indentation in the natural landscape is used or a pit is actually dug for this purpose. This type of disposal has the advantage of relative ease of construction and disposal and, as a result, it is also the cheapest way to dispose of waste. For this reason, it remains by far the most prevalent way in which waste is disposed of in contemporary cities of the Global South.

The main environmental problems associated with open-pit dumping include the already noted attraction of disease-bearing rodents and insects, as well as the openly accessible emergence of toxic microbes. Also, with time and precipitation, toxic **leachate** from open-pit dumps can pollute both ground and surface water supplies for the greater city population. This latter has been particularly the case over the last 150 years or so as the composition of solid waste from the city has changed from dominantly organic to dominantly inorganic materials. From things such as animal matter, food scraps, wood, plain paper, and other more or less biodegradable materials, such waste has come to consist increasingly of things such as dyed clothing, plastics, rubber tires, paints, batteries, scrap chemicals and chemically treated packaging, plastics, and, most recently, **electronic waste** (e.g. insulated wiring, silicon, and picture tubes).

Not only does such an increased amount of inorganic waste take much longer to degrade naturally, it also leaches many toxic chemicals once this process begins. Not only that, but in poorer countries the ever-growing numbers of city trash pickers/recyclers come into direct physical contact with such materials, which has led to increasing skin and respiratory diseases among them.

This broad historical trend is extremely important to keep in mind. As countries in the Global South continue to develop economically, for example, city populations have increasingly adopted consumption patterns quite similar to those of their Northern counterparts. Everything from fast food in disposable containers, modern medicines and industrial chemicals, processed and elaborately packaged commodities, appliances, on to automobiles, cell phones, iPods, and computers are increasingly being consumed by Southern elites. In this sense, as disposable income has increased for this minority of city people, the consumption of such rapidly disposable, predominantly inorganic commodities has increased in turn. This signals the beginning of a truly global problem of city-generated solid waste to be sure.

In the Global North this trend in the production of waste has been met largely by innovations in disposal, particularly in terms of more sanitary burying in landfills as well as burning. Landfilling solid waste entails artificially and intensely compacting such waste with heavy machinery and covering such dumps with layers of soil. This process allows for a greater quantity of garbage to be dumped in any given space and the soil coverage usually prevents the long-term infestation of vermin and the escape of disease-carrying microbes. Also, if engineered properly, the bottom, and often even separate layers, of landfills are lined to prevent the percolation of toxic leachate, thereby protecting nearby water and soil resources. Unfortunately, all of this effort to dump city waste in a more sanitary manner costs money, thereby putting such efforts out of the reach of most cities in poorer countries and of those relatively richer cities now experiencing fiscal crisis.

Moreover, given the ever-greater quantities of garbage, in general, and especially inorganic waste generated by city people in richer countries, these advantages of landfills are not always realized. Such dumps also tend to produce and sometimes emit high concentrations of methane gas, which is highly toxic and contributes significantly to the local atmospheric ozone problem in major cities. Furthermore, as cities have grown, more and more territory has had to be used for this purpose and the trend has been toward regionalizing landfills farther outside the boundaries of major cities. This necessitates transporting ever-greater quantities of city waste ever-greater distances, utilizing ever-greater amounts of energy in the dumping process and contributing to the air pollution of metropolitan environments, in general.

Because of these issues, many cities have taken to burning more and more of their waste. Incineration has two major advantages. First, it gets rid of much solid waste altogether and, second, the very process of burning creates a possible source of energy. However, burning solid waste has potentially quite significant implications for air quality, particularly given the relatively high levels of inorganic, chemically laden solid waste, such as rubber and plastic ware, paints, chemically treated paper, and so on, generated in richer cities. Even in poorer cities where such incineration takes place, particulate matter is emitted into the air, which is a significant irritant and potential long-term health issue for city dwellers, particularly those already with respiratory problems as well as the very old and the very young.

CATCHING YOUR BREATH IN THE CITY

These last points signal the next major aspect of city environments that needs to be discussed: air quality. With so many people living and working in such close physical proximity, air quality suffers as a result of everything from smoke from cooking and heating to emissions from artisanal workshops and industrial factories to windswept dust from unpaved roads and on-going construction sites. Such air pollutants are concentrated and, because of changes in surface cover as well as altered wind patterns and other microclimatic changes resulting from city life, noted above, tend to hang around cities instead of drifting out. This issue of city air became quite obvious during the Industrial Revolution as more and more coal was burned for heat and cooking, as well as for industrial production. Industrial cities literally became lost in their own smog and their built environments (to say nothing of the lungs of city people) became literally covered with coal residue. Indeed, one can get a sense of the nature of such air pollution in the slums of many post-colonial cities today where unregulated burning of particularly charcoal fuels many a household and informal workshop.

Today, city air is fouled by all manner of emissions, from factories to, most importantly, automobile and truck exhausts. From dense central city streets clogged with congested, idling internal combustion engines to ever-sprawling yet also increasingly congested suburban and even exurban streets, city air is increasingly contaminated by the very automobility that has marked city development over the last century. And this promises to be even more the case globally as poorer cities become richer and more and more people living in them come to own cars and trucks.

More generally, city air pollution can be characterized in terms of primary and secondary pollutants, as well as airborne particulates. Primary air pollutants are emitted directly into the air and include carbon monoxide and non-methane hydrocarbons, including propane, propene, and acetylene. All of these emissions come from the incomplete burning of fossil fuels and can reach toxic levels as they concentrate in cities. Another potentially toxic emission related to the burning of fossil fuels is sulfur, which not only affects human respiratory function but also, when mixed with moisture in the air, produces what has been called **acid precipitation**, which destroys vegetation and even damages the built environment of cities.

Secondary air pollutants are emissions that can become toxic when mixed in this way with atmospheric conditions. An example would be the mixture of chemically laden particulate ash from the incineration of solid waste with moisture in the air to produce acid-like precipitation. Particulate matter also gets stirred into the air of cities from urban soil and road dirt, as well as from the construction and destruction of the built environment. The most lethal of such particulate matter for city people involves that which consists of lead, which is still present in the gasoline used to fuel cars and trucks in many countries in the Global South as well as in many industrial and household paints still in use. Of most concern, even in those countries which have banned leaded gasoline and paint, lead still lingers in the paint on older buildings and in city soils.

ENVIRONMENTAL JUSTICE AND THE CITY

Of course, this has been only a very brief overview of the environmental impacts of cities and city life. Of most importance is that cities represent an extreme intensification of the human modification and, in many instances, degradation of non-human nature, both in their immediate hinterlands and now quite globally. Cities of the Global North continue their sprawl of citified bits over ever-greater areas of formerly rural territory. In turn, the unregulated, ever-expanding suburban slums of Southern megacities continue to gobble up more and more hinterland. As a result, the degradation of global nature on the part of city people seems destined to expand and intensify even more in the near future unless such trends are somehow attenuated. Unfortunately, this seems to be more and more impossible, given the increasing lack of knowledge of natural processes among Northern city people and lack of other opportunities for mere survival among the ever-growing numbers of Southern slum-dwellers.

This, unfortunately, appears particularly the case when the justice of this situation is considered. In relatively rich Northern cities, the per capita use of the Earth's natural resources is such that their resource "hinterland" is now quite global in extent. Indeed, many argue that there are simply not enough natural resources for all people of the world to attain such levels of material riches, given the way in which such resources are used in the Global North. If that is true, and if it is also true that richer city people in the Global South increasingly are adopting the material lifestyle of their counterparts in the North, as discussed above, this certainly does not bode well for the future, either for non-human nature or, indeed, for those humans much less able to procure such a livelihood. In other words, nowhere near all Brazilians or Russians or Indians or Chinese or South Africans (the so-called developing "middle-class" **BRICS** countries) can even hope to attain the level of material wealth of most city people in the Global North. The natural **carrying capacity** of the Earth simply cannot support such levels of energy use and material consumption and, at the same time, absorb such levels of city waste that would be entailed. The fair share of such carrying capacity for people of poorer countries, in other words, has already been used or otherwise degraded by those in richer ones.

In fact, this environmental injustice on a global scale can be found in microcosmic form in the very duality of most cities today. As discussed in Chapters 4 and 5, de-industrialization, off-shoring, and the suburbanization of richer city people in the Global North has left most inner cities in sad material shape, both in built and social terms. People stuck in such inner cities because of lack of opportunity now live disproportionately near former industrial, now heavily polluted, **brownfield** sites. In addition to the health implications of this situation, in other words, the natural resources of inner cities have already been used and degraded and are now unavailable to the local population as a means for material advancement.

Similarly, the slum-dwellers of the Global South live disproportionately in marginal areas of the city, either former, similarly polluted, industrial or transportation sites, around open-pit garbage pits, along already polluted intra-city waterways, or precariously perched on ever-crowded pieces of land that cannot possibly accommodate and absorb their patterns of consumption and waste. At the same time, their richer city neighbors increasingly enjoy the resource-intensive material lifestyle of suburban city people of the Global North. In the end, if one is to understand the environmental impact of cities and city people in today's globalizing world, it is necessary to pay close attention to such environmental justice issues, both at the global and city scale. This is particularly the case because, as the global population becomes ever-more majority urban, such issues, if not adequately addressed, are likely to result in no little social conflict both within cities themselves and, indeed, on an increasingly global scale.

Box 8 The city and gender

Regardless of the liberating potential of city life, women today still mostly find themselves to be lower paid for the same work and otherwise relegated to pink-collar jobs as secretaries and other lower-status occupations. Indeed, the history of industrial cities shows this continuing patriarchy in city labor markets quite explicitly. It did not take long into the industrial period for women (and children) to be forced out of factory work by labor laws and community persuasion. Women thus became privatized in the household and were increasingly understood in the new Victorian ideology of gender to be the "weaker sex" (in mind and body), not fit for a public role outside the home even to vote in popular elections.

As industrialization continued, then, most women were increasingly relegated to the privatized times and spaces of the city, and this is where the contemporary city process of mass suburbanization becomes important not only for environmental, but also for gender, issues. As private and public time spaces of the city emerged and were reinforced in an increasingly patriarchal manner, over time the separation between the two became ever greater and deeper. Whereas, at first, women were privatized in their city households, these households at least were still centrally located in the city in dense agglomeration. So women continued to constitute a major proportion of city populations and were thus more able to make themselves apparent in "public" places and to form, and make heard, socially significant "public" voices in community with other women in close spatial proximity.

Mass suburbanization, however, changed all this. With suburbanization came the privatization of women most dramatically in detached single-family houses far from the densely packed social relations of the city. This increasing spatial separation of the public and private space times of industrial cities simply reinforced, and quite clearly deepened, the privatization of women. Now women not only were in "living spaces" far away from any viable "working spaces," they were also increasingly alienated from other women alone, as they were each working day, in their separate house surrounded more with household gadgets than other people. Now patriarchy was etched quite clearly and visibly in the very social and built environment of the sprawling industrial city.

Yet, again, these socially and materially constructed gender relations in the city may be undergoing significant change as a result of three more recent post-industrial processes. First, the real wages of most workers have stagnated since the 1970s. This has necessitated two worker families to maintain standards of living and this, in turn, has meant that more women have entered the public labor market. Second, the rise and social successes of the feminist movement during this same period forcibly created more public voice and space for women in the city, thereby reinforcing the first trend. Finally, the suburbanization of business in the post-industrial metropolitanization of city economies has resulted in a much more diversified labor market on the outer fringes of cities. As a result, more high-paid professional jobs are available in suburbs and exurbs and these can be filled by women as well as men.

In short, these new post-industrial processes appear to be shaking up gender relations in the city once more. It just may be that, in the case of women, they will be as liberating as those of the early industrialization process. But it may also be that this remains only a hope that needs continual social activism for it to become a reality. In any case, for those concerned with gender and the city, a closer examination of what is right now taking place, both in post-industrial cities as well as in newly industrializing cities, will certainly provide clues to what the future may hold for the social construction of gender in the city of the future.

FURTHER READING

A close examination of environmental justice issues applicable to cities can be found in Robert Bullard's *Dumping in Dixie: Race, Class, and Environmental Quality* (Boulder, CO: Western Publishing, 2000). A more intensive look at the alteration of nature as a result of city development can be found in Matthew Gandy's *Concrete and Clay: Reworking Nature in New York City* (Cambridge, MA: MIT Press, 2002). Lauren Heberle and Susan Opp take on a more global focus in their edited book *Local Sustainable Urban Development in a Globalized World* (Burlington, VT: Ashgate Publishing Company, 2008), while Maria Kaika describes how nature and city life merge into socio-natural phenomena in her extremely interesting book *Spaces of Flows: Modernity, Nature, and the City* (London: Routledge, 2004). More generally, Anthony Penna provides a fine overview of the notion of ecological footprints in his *The Human Footprint: A Global Environmental History* (Malden, MA: Blackwell Publishers, 2010). Yvonne Rydin lays out the planning and policy aspects of environmental sustainability in her *The Purpose of Planning: Creating Sustainable Towns and Cities* (Bristol, UK: The Policy Press, 2011), and Stephen M. Wheeler and Timothy Beatley provide a comprehensive overview of this issue in their edited volume *The Sustainable Urban Development Reader* (New York, NY: Routledge, 2004). And Lynda Johnston and Robyn Longhurst have produced a very comprehensive book on gender and geography: *Space, Place, and Sex: Geographies of Sexualities* (Lanham, MD: Rowman and Littlefield, 2010), which also speaks to the city experience of gender relations.

9

CITY PLANNING

Since the beginning of city life, the built and social morphology of cities has been planned, at least in terms of areal and social focal points. Until the Industrial Revolution (ca. 1750), this generally meant that the very central core of cities was planned to materially display religious, political, or mercantile power, as well as to house those city dwellers who most wielded such power. As cities grew in population, material displays of power such as statues of city elites, memorial fountains, war victory memorials, military, police, and justice buildings, etc., were consciously located outside this center at strategic, usually most visible, locations within the city by the powers that be. Yet, for the most part, city elites themselves remained within the most central parts of the city and these parts remained the most actively planned in terms of built structures, housing distribution and occupancy, and overall street patterns. In short, city planning before the Industrial Revolution was a partial affair, covering only a small portion of the city and catering to the desires and needs of city elites alone, much like all "public works" did in such cities, as discussed in the previous chapter.

As the populations of such cities grew, the outer rings of their built and social morphology were established more or less helter-skelter by new city people without any sort of planning whatsoever other than, in most cases, some notion of the actual territorial boundary of the city. That this latter was usually a wall of some sort for protection against military attack from outside the city meant, in turn, that as such cities grew in population, their built and social morphologies became ever-more densely packed. Of particular note, save for the walled enclosure, post-colonial megacities today provide many an example of such partially planned, mostly organically grown cities as their sprawling suburbs continue to expand with ever-greater numbers of recent immigrants from the countryside. These poor suburbs, just like those in pre-industrial cities, generally are not supplied with adequate public works infrastructure and the people who inhabit them must meet their housing needs as best as they can by themselves.

In more specific terms, pre-industrial city planning focused on what is known as **monumentalism** in the sense of exhibiting social power by creating relatively huge built monuments that announce such power, both by their very size as well as by their elaborate architectural embellishment. From the palaces of emperors, caesars, kings, and the richest merchants, to monuments of war achievement and police power, on to monuments to specific city elite themselves, such edifices within the city were intended not only to announce social power within the city. They also, by their very imposing and obvious presence, were meant to maintain such power by eliciting awe, if not fear, in the minds of less powerful city dwellers.

Nevertheless, even if early city planning was only partial in this respect, later city planners learned from such early planning everything from more efficient street and central square planning, essential water and sewer provision, and the very role that built structures play in the maintenance of social order in the city. In turn, the study of older built monuments in cities has yielded insights into the proper engineering of such structures. It also has provided many an inspiration for architectural innovation. Indeed, the very idea of the conscious planning of city life, however partial and socially oppressive these early attempts may have been, is already a social innovation that should be acknowledged.

NEO-MONUMENTALISM AND THE BIRTH OF MODERN CITY PLANNING

As discussed, the Industrial Revolution brought dramatic changes to city life itself as poorer, much less powerful city people came to dominate central city populations in ever-denser built and social clusters. In this respect, the earlier planned areas of the city for the elite, to the extent that they existed, were literally overrun and largely remodeled according to the needs of newly dominant economic activities and other city people. Such remodeling entailed ever-denser built structures, ever-narrower streets and alleyways, and an ever-growing need to make use of every bit of open space, including city parks and other public squares, to house expanding city populations.

Again, this major transition in city life brought with it all sorts of economic, political, cultural, and environmental problems. Here, however, the emphasis is twofold. First, the rapidly growing presence of this new non-elite majority in central cities posed a much greater threat to city elites than in pre-industrial cities. Whereas in early cities, and in most post-colonial cities today, the non-elite majority was largely controlled by planned and enforced physical segregation within the city, this new city majority literally took over major centrally located areas of the city. Because of the rapidity by which this transition took place, many city elite, as a result, were literally surrounded by non-elite city people and their needs.

Second, the growth of industrial cities was based essentially on the decisions made by private individuals in private property markets. That is, there was generally no overall plan for industrial city layout and growth as at least partially was the case in pre-industrial cities. Nor were the people behind such growth in industrial cities concerned with such. Even in merchant cities, for example, many richer merchants took on a public role in planning the cities they dominated. In industrial cities, however, major industrialists and city property developers did not have to consider the public effects of their decision-making as to where to place the factory, how to build it, what kind and, indeed, quality of housing should be built for new factory workers and where, and so on. Some of them did get involved in city politics to be sure; but, importantly, this was not considered a requirement.

As a result of these two realities, industrial cities grew quite organically with very little planning whatsoever – except, that is, the retroactive planning that became absolutely necessary in order to combat the very public problems that the industrialization process itself produced. And this, then, is the key. As discussed in previous chapters, problems such as maintaining social order, providing clean water and efficient sewerage to prevent infectious diseases, and preventing widespread fire damage are endemic to the densely packed populations of cities in general. With the rapid urbanization brought on by industrialization, such city problems were very swiftly intensified and began unavoidably to affect all city people, elite and non-elite alike, throughout the city. Because of the rapid influx of mostly destitute and desperate rural people to the city, for example, traditional policing and justice systems were soon overwhelmed by new demands. Similarly, infectious diseases which originated most often and spread most quickly in the poorest neighborhoods because of inadequate water and sewer systems could no longer be contained within them. Finally, because of the rapidity by which new city housing was built and at what density, fires spread much more quickly than organized fire-fighting could handle. In short, these fast-emerging problems instigated by the needs of the growing numbers of city people "below," rendered it more and more impossible for city elite to avoid or otherwise escape such city-bred social and environmental problems.

THE HAUSSMANNIZATION OF PARIS

Modern city planning, then, was a socially forced affair which eventually emerged in all industrializing cities by necessity. The best example of this emergence, however, took place in Paris during the time of Napoleon III, specifically under the direction of Georges-Eugène Haussmann. Haussmann set out to completely remodel the city of Paris between 1853 and 1870 to better meet the growing social and environmental needs of the industrializing city. At the time, most of the city's public works infrastructure, street, and building patterns remained in the organic arrangement of earlier times and had been quite overwhelmed by an ever-growing population. As a result, Paris was mostly overpopulated, congested, filthy, and crime and disease ridden. Equally important from Napoleon's point of view, the city was quite unstable socially, having been the site of several city riots already which threatened not only city, but also national, authorities, given its role as the capital city of France.

Under Napoleon's direction, Haussmann set out to render the city more physically "efficient" by creating what was considered to be a more modern street network that literally crisscrossed the city. This new street pattern, it was thought, would result in a more efficient traffic flow for business purposes. It also would make it much easier for the military and police to access the entire city to enforce social order, particularly in the majority overpopulated slum areas. These new streets were constructed in grand boulevard fashion, much wider and straighter than other city streets and lined with newly planted trees and wide sidewalks. They ran right through the poorest parts of the city, demolishing much housing, however inadequate, of the most marginal of the city's population. In this respect, Haussmann's plan was an early version of the combined intra-urban highway/urban renewal policies of the United States in the late twentieth century, only with even less regard for the now homeless slum dwellers of Paris.

In many ways, Haussmann's grand plan for the remodeling of the built environment of Paris in the nineteenth century was quite similar to the earlier monumental city planning of Ancient/Classical cities. Indeed, the fact that Haussmann constructed grand central plazas with war and other memorial centerpieces suggested no little emulation of such earlier city plans. Similarly, the Haussmannization of Paris was a top-down affair, ordered and directed by Napoleon III and implemented by Haussmann with very little input even from city elite. Nevertheless, it can be considered the beginning of modern city planning in two ways. First, the clear intent was to plan on a much larger, close to city-wide, scale and not just for the elite districts of the city. That new sewer and water lines were constructed under the new boulevards also suggests a more city-wide and even future-oriented planning process.

Second, the fact that these problems were endemic to industrializing cities everywhere meant that Haussmann's plan for Paris was closely watched by authorities in other rapidly industrializing cities and was soon emulated in cities such as Lyon, Marseilles, Brussels, Birmingham, and even Rome. It was clear in these new city times that the non-elite city populations, now overwhelmingly in the majority and ever growing in numbers, could not be controlled simply by forced segregation or the awesomeness of memorial monuments scattered around the city. In short, while the planning and implementation of Haussmann's plan for Paris may have been a top-down affair and included no little monumentalism of its own, the plan itself was directed at solving the quite modern city problems brought on by industrialization itself.

RENDERING THE "CITY BEAUTIFUL"

Indeed, the specific characteristics of **Haussmannism** provided a longstanding, much emulated, model for future city planning both in industrial and colonial cities. The example set by Haussmann, for example, was a conscious inspiration for Robert Moses in his attempt to remake New York City big time in the mid-twentieth century. In general, Haussmann's influence can rather easily be discerned in terms of modern city planning's bias toward the built environment of cities as well as its relative neglect, until most recently, of the needs of the majority of non-elite city dwellers. City planners generally have been notorious environmental determinists in the sense of believing that social problems can be solved by manipulating the built environment. As discussed below, this **environmental determinism** continues up to the present day in the form of much post-industrial, post-modern city planning. In turn, Haussmann's attempt to solve the problem of the seriously over-crowded, disease-ridden, increasingly socially restless slums of Paris was not meant to provide a more equitable socio-economic solution for the majority of the city's population who lived in them. Rather, it was to physically destroy such neighborhoods and, at the same time, render the surrounding neighborhoods more socially passive via the new ease of wide boulevard access to them on the part of the police and military. That the value of the property which now fronted these new boulevards rose dramatically also ensured that only the city elite would be accommodated by any new housing, which resulted in the slum clearance process.

Modern city planning has largely continued this bias toward the manipulation of the built environment and the relative neglect of the underlying social and environmental causes of city problems. In the United States, for example, early city planning focused mainly on the creation of parks and other built amenities in order to render cities more pleasing to the eye. Haussmann's plan for Paris, again, was an early model in that, in addition to creating the great boulevards which pierced through the old city, several large open-access parks were created. In the United States, cities were growing at a rapid rate at the turn of the twentieth century, particularly fueled by foreign immigration. As a result, slums of inadequate housing grew in size and extent and most of the other problems of industrial cities, in general, emerged very quickly and quite dramatically. Faced with this situation, many authorities began to look to the creation of city parks as a means of alleviating such social and environmental problems.

This idea of nature as a cure for social ills also runs through the modern history of planning. In fact, it follows from the general belief that city problems are the result of city life, in general, instead of how cities are developed and by whom. A more rural life was considered more socially safe and environmentally healthy. And streetcar, and eventually automobile, suburbs and now exurbs were, and still are, developed on the basis of this sort of anti-urban bias. This idea of a natural cure for social ills is also, of course, what was behind the perceived necessity of creating grand city parks. City social problems would be solved, according to this view, by having large open and adequately policed green spaces where city elites and non-elites could mingle together to achieve greater social harmony. In turn, large urban parks were thought to provide access to a cleansing nature where bad disease-carrying city air could be expunged and good healthy and natural air could be inhaled. New York City's Central Park, Brooklyn's Prospect Park, and Buffalo's and Boston's stringed park systems thus were meant to solve the social and environmental ills of over-urbanizing industrial cities simply by being built.

Such assumed ameliorating characteristics of nature thereby became a staple of modern city planning. Of most importance, however, is that the parks movement in the United States was really part of a more general planning movement focused on rendering cities more aesthetically pleasing by creating, in addition to parks, new civic architecture and symbolic buildings, civic art projects, and general landscaping along central streets and in open spaces. This wider movement to render the **city beautiful** was a result of activities by city improvement associations and clubs throughout the country which were organized and run by middle- and upper-class women. Such associations sought to clean up city neighborhoods via street and open-area sweeping and garbage collecting by volunteers and other public relations campaigns to render cities more livable. These early volunteer efforts were then adopted by city planning agencies on a much wider scale.

The new city beautiful movement concentrated, like Haussmann and the parks movement, on the built environment. There was much emphasis on the symbolism of the architecture of civically important public buildings such as city halls, police and fire stations, libraries, performing arts centers, schools, railroad stations, and history and art museums. Also, great centrally focused and tree-lined boulevards were carved into cities and open spaces garnished with civically important statuary. Of significance is that this was a planning movement that sought to create a better image of the entire city mostly to attract and maintain city elites and businesses. In this respect, the city beautiful planning movement of the early twentieth century in the United States was not much different from today's city planners, who similarly attempt to beautify and market their post-industrial cities better to attract increasingly footloose business firms and more elite city people. In both cases, the attention of planners and others is drawn away from the true causes of city problems as focus is put singularly on the built environment as a means to improve city life.

MAKING PLANS TO HOUSE THE CITY MASSES

In many respects, the city beautiful movement in the United States borrowed from the neo-monumentalism of post-Haussmann European planners and also diffused its own ideas of a grander scale and extent of civic embellishment back to Europe. But, of course, the social and environmental problems of industrializing cities were not solved by such planning efforts as much as simply redistributed to other parts of the city. More or less every time city land was taken for each new boulevard or memorial square or park or civic building in the process of beautification, for example, less and less city property was available to house the ever-growing numbers of non-elite city people. As was the case with much more recent urban renewal policies in the United States, this put enormous pressure on the housing stock of non-"beautified" areas of the city as demand for housing increased as a result.

This increase in demand for available housing simply exacerbated the housing problem in cities as rents went up and already shoddy housing units became even more crowded. Eventually, this growing problem of inadequate housing for the masses within cities could not be ignored, particularly as increasing numbers of reports concerning the living conditions of slum dwellers in major industrializing cities were published. At first, there were private volunteer attempts to ameliorate this situation as philanthropists constructed early versions of public housing and health workers attempted to provide personal assistance to slum-dwellers. However, given the growing magnitude of the problem in most industrializing cities, these efforts soon proved quite inadequate and local and extra-local government agencies were forced to step in to determine how best to solve such a growing public problem.

City housing for the masses thereby became another impetus to modern city planning. In Europe, the bulk of such housing eventually was constructed and managed on the basis of public spending and management. This became particularly the case as housing in cities needed to be reconstructed on a massive scale after the devastation wrought by World War II. Because of this, former slum neighborhoods took on a more uniform physical appearance and most of the urban poor were able to procure more or less adequate housing at rents they could afford. Of importance also is that the new public responsibility for city housing rendered the conscious planning of cities that much more institutionalized in Europe. This simply reinforced the traditionally strong government planning role in the political economies of European countries. Moreover, this extension of public planning in the city signified the beginning of the process by which the more consciously planned European cities set off on a development trajectory quite different from that of the ever-sprawling cities in the United States, as discussed in Chapter 6.

Indeed, until as late as the 1960s, the supply and condition of housing for the lower classes in U.S. cities was left to the private market of property owners and developers. There were early experiments with public housing in the 1930s as part of national policies to combat the Great Depression. For the most part, however, what planning there was in this important area of city development in the United States consisted essentially of land-use zoning and formal, yet unevenly enforced, building codes. Particularly because there was no wartime damage on the scale that existed in Europe, the housing needs of the majority of city dwellers and returning World War II veterans were not considered by most city authorities to be central to the overall development of cities. Such needs, it was generally thought, would be taken care of via decisions made in private housing markets. The results, of course, were highly uneven as overcrowded slum-like conditions continued to exist in most major U.S. cities due to both inadequate housing stock and active discrimination in city housing markets. In turn, the built environments of cities continued to sprawl outward toward the suburban fringe with little conscious planning at all and especially not with regard to the housing needs of less fortunate city people.

Nevertheless, the resulting deterioration of central cities and widespread urban unrest and rioting during the 1960s eventually forced some recognition of the need for more active city planning, in general, and with regard to housing specifically. Unlike previous city beautiful efforts, the combined and quite major effects of the federally funded construction of intra-urban highways and resulting destruction of entire neighborhoods of the poor led many to argue that replacement housing was absolutely necessary for the good of the city as a whole. New centrally located, densely clustered public housing projects which can still be seen in many U.S. cities beside intra-urban highways were the result, as detailed in Chapter 6.

MORE RADICAL PLANS FOR THE CITY

The history of city planning also includes more radical plans for solving the social and environmental ills of industrializing cities, at least as initially conceived. An example of such planning was enunciated most directly and comprehensively by an Englishman, Ebenezer Howard (1850 to 1928), who called for the development of what he called *garden cities*. Garden cities were to be fully planned not just in terms of their built environments but also in terms of social relations. The population of each city was to be kept to a certain size, about 30,000, housing and businesses were to be adequately spaced and sized, landscapes were planned with natural green settings, and, as importantly, city property was to be owned cooperatively among all the citizens of the city. Garden cities thus were to be fully master planned, quite literally from the ground up on greenfield sites, and, in this sense, were to be distinct alternatives to cities as they were then developing.

The idea was to get people of all different financial means out of unhealthy, congested industrializing cities and to start building city communities anew in much smaller urban clusters which would combine human constructions and nature equally. Significantly, garden cities were to include housing and business and therefore be relatively self-contained living and working sites instead of being simply bedroom suburban communities tied to larger city labor markets. Also significant, of course, is the idea of the communal ownership of land. The idea was for potential garden city inhabitants to pool their money to leverage additional loans in order to buy adequately sized parcels of relatively cheap agricultural land. Then, as the cities were built and inhabited, inhabitants eventually would profit from the increasing property values that accrued simply from such property improvement and something like the bid-rent curve described in Chapter 5. With this profit, garden city residents would be able to pay off their original loans and, most importantly, begin to accumulate a social fund for the city to be used for social services for the benefit of all residents.

Now this was obviously a radical idea. As such, it was a clear alternative to the uneven, relatively authoritarian, city planning of the neo-monumentalism of Haussmannism and the city beautiful movements. It was also more comprehensive city planning in that it focused on the entire city, including both the built and social environment. The notion of cooperative ownership and management of city assets was particularly crucial from Howard's point of view in that it would maintain cooperative "ownership" and therefore care among inhabitants for not only the built structures of such cities, but also the human community that such cooperation demands. To the extent that this was the case, in other words, it was thought that the environmental and social problems of traditional industrial cities simply would not arise.

In the end, the garden city ideal arguably was made a reality in only two cities in England, Letchworth (founded in 1903) and

Welwyn Garden City (founded in 1920), and then only partially so. The built environments of these cities generally fit the model of better blending human constructions with nature. They both also included residents and their businesses, and the initial founding was based on cooperative ownership. But over time, these cities have lost their cooperative ownership characteristics and have largely become suburbs to larger cities in their vicinity instead of self-contained parts of a network of such garden cities linked together by public transportation. Indeed, this latter part of Howard's plan generally has been ignored in all later attempts to follow, in some sense, his garden city ideal. Howard was not seeking to create small rural towns simply as a nostalgic alternative to big industrial cities, as is mostly the case in today's post-industrial city planning. Rather, master-planned garden cities were to replace such overcrowded, congested cities by redistributing their populations to more manageably sized built and social environments surrounded by greenbelts and yet linked together as metropolitan regions via public transportation modes and networks. In this, then, garden cities could very well accommodate the rapidly growing numbers of city people resulting from industrialization yet in a much more environmentally and socially sustainable manner. This is another way in which Howard's garden city ideal was radical and, indeed, much ahead of his time.

Just like Haussmannism, the garden city conception has been enormously influential in the history of city planning. As discussed below, something like Howard's plans have been behind most every experiment in master-planning cities or districts therein since his time, from the **new town** movement after World War II to more recent suburban subdivisions on to the so-called **new urban** experiments in post-industrial, anti-automobile town planning today. In the United States, early adherents to garden city ideals formed the Regional Planning Association of America (RPAA), which had its hand in many a planning venture in the early twentieth century, particularly with the pre-Depression plan for the city of Radburn, New Jersey. Nevertheless, in virtually all of these latter-day planning ventures, the very radical nature of the garden city ideal became severely attenuated, although the RPAA at least understood that the garden city ideal was not about nostalgia for small town life *per se* but, rather, a serious alternative to unsustainable industrial over-urbanization. In any case, the ideal of cooperative ownership and management, the reliance on public transportation, and the extent to which cooperative public planning would have had to be maintained in order to realize the garden city ideal more or less doomed it from full realization in private market-driven countries.

PLANNING THE POST-WAR "NEW TOWN"

In many more conventional ways, the garden city ideal followed beliefs already behind the design of streetcar suburbs and the urban parks movement. This is particularly the case with regard to the perceived necessity of allowing city dwellers more contact with nature in order to help solve the social and environmental ills of city life. Indeed, this focus on nature unfortunately has been the main legacy of the garden city ideal. Instead of attempting to tackle the true causes of city problems, as Howard had originally intended, what have materialized on the basis of this legacy are more or less garden suburbs, master planned by private developers with all the "natural" and leisurely demands of more wealthy city people in mind and usually enclosed behind guarded gates, as described in Chapter 7.

But there was another attempt at new city planning after World War II that mimicked the garden city ideal in a less privatized manner. This was the new town movement which originated most clearly in the United Kingdom with the passage of the 1946 New Towns Act. As already noted, the war took a great toll on the built and social environments of European cities. Post-war planning in most countries thus entailed much rebuilding in city centers and, significantly, attempts both to limit their size via enforced **green belt** boundaries and to disperse city populations out to smaller satellite towns. Around London, for example, early new towns were planned around older villages such as Stevenage, Harlow, and Hatfield, and several others. In turn, the new towns of Glenrothes and Cwmbran arose in Scotland and Wales, respectively (1940s to 1950s). Later English new towns were created in places such as Milton Keynes and Peterborough in England and Irvine in Scotland (1960s to 1970s). Similar towns were planned and developed around Paris and other parts of France and, indeed, eventually throughout Western Europe and even in the United States in places such as Reston (Virginia) and Columbia (Maryland).

Other than the stated planning goals of limiting the size and bounding such new towns with green belt moats, however, it is difficult to characterize precisely what was behind, and even included in, the new town ideal. Some argue that to be a true new town it needs to be constructed from the ground up on a truly greenfield site. And some towns were built in this way, such as Louvain-la-Neuve in Belgium and, perhaps most famously, the new capital city of Brasilia in Brazil. Others, however, argue that new towns are simply better, more extensively planned, extensions of cities that already exist. Still others argue that such towns are intended to be more socially innovative, particularly by allowing for ethnic and income diversity and housing style. This lack of precise definition is probably based on the fact that most "new town" planning has focused largely on the built environment as a means to shape city life. And, in turn, the very ambiguity of the notion provides wide latitude for planners to announce their projects as something that is, indeed, new and therefore innovative. No wonder, then, that planned "new towns" have been announced now virtually around the world, from Asia to the Middle East to Africa and the Americas, as any simple internet search will reveal.

In any case, aspects of new town planning persist in most city planning efforts today. This is particularly the case in terms of the various concerted efforts to limit sprawl in the so-called **smart growth** movement in the United States. In turn, the new town ideal of green belting cities plays a large role in both smart growth planning as well as in the plans of those who are now pushing for more environmentally sustainable city growth. Finally, something like the new town ideal is behind the many recent, privately planned suburban, and, especially, new urban communities which now are attempting to attract business as well as residents. In the end, the new town ideal is one which lingers both because it remains ill defined, and therefore open to much interpretation, but also, arguably, because it entails the notion attractive to most professional city planners of master-planning an entire city in order to finally get things right.

TOWARD POST-MODERN CITY PLANNING?

The post-World War II period really marked the height of city planning, particularly in rebuilding Europe. This was a time when much simply had to be done to get countries and their major cities back to working order. The fact that European countries have traditionally stronger, more centralized public planning, in general, simply meant that big city plans, such as those resulting in whole new towns, proliferated. By the 1960s, however, there were already signs that many city people were not pleased with such big plans or the way in which they were being carried out. Even in the United States, where city planning has always been less powerful than private interests in terms of shaping city life, there was public dissent over the way in which urban renewal and infrastructure development was taking place. To characterize such dissent broadly, many on both sides of the Atlantic began to decry the top-down, authoritarian way most city planning was undertaken. Such critics also decried the very material and social effects that such planning was having on cities. Whole traditional neighborhoods and towns were being fully remodeled in mostly mass-produced, uniform ways. And fully master-planned "new towns" seemed, similarly, to be lifeless, with little diversity or aspects of truly human tradition or interest in their instantly produced, all-enveloping built environments.

Indeed, much like how Haussmann attempted to render Paris more efficient and functional for the new industrial age, post-war city planning seemed to be attempting to render cities simply more efficient and rational for the new increasingly post-industrial economy. Office space had to be provided and automobility ensured. A new town such as Milton Keynes in England was fully designed for such and its drab, simply functional grid-street pattern and most of its built environment reflect this. Even the more radically conceived new town of Reston, Virginia, came to participate, and essentially be engulfed, in the post-industrial metropolitan sprawl of Washington, D.C. All cities were beginning to look alike, from extensive intra-urban highway corridors which now acted as spatial barriers within the city to the high-rise, no-frills, glass-encased office towers that dominated city horizons. In short, many began to suggest that cities were being planned not for people, but solely for economic efficiency and functionality as well as automobility.

Similarly, criticisms of lifelessness were being made concerning the most extensively planned new towns such as Brasilia and Louvain-la-Neuve. The built environments of such towns had certainly been constructed from the ground up according to the best social theories held by their planners. But the social results from such a full manipulation of the built environment by these few professionals did not turn out as they planned. Instead, many new residents rebelled against the drabness and uniformity of such total planning of the environment in much the same way as did the new residents of similarly drab high-rise public housing projects in poverty-ridden inner cities in the United States. In both cases, in other words, the fully planned built environment of the place elicited quite the opposite reaction among residents than anticipated by planners.

As a result, many began to push for changes in the way in which cities were being planned. Some argued, for example, that there should be greater participation of actual city people of all walks of life in the planning process. Such **collaborative planning** would ensure, the argument goes, that cities are planned with the true interests and desires of residents in mind from the beginning. Such planning would involve discussions among important city-based stakeholder groups, from business representatives to activist groups to average citizens in order to ensure full representation of all such interests and desires in city plans. Others contended that more emphasis in city planning needed to be put on preserving the unique historical and cultural legacies that had already evolved in cities. Instead of mowing down all old or culturally unique structures within the city in order to build the next no-frills, box-like, glass-encased office building or apartment/condominium complex or, indeed, intra-urban highway spike, this argument goes, more concern should be placed on preserving such built diversity. In short, the **historical preservation** of city neighborhoods is important, according to this view, because such neighborhoods reflect a collectively built environment and therefore the culture of the people who built it and continue to take pride in it, at least in theory.

In a very real sense, these new movements within city planning reflect a post-modern attitude to the process by suggesting that, instead of entailing simply professionally conceived, top-down implemented "big plans" for efficiency and functionality, city planning should include many more voices in conception and therefore allow greater diversity in result. To the extent that it does, the more all city people will take greater possession of their city and therefore pride in its upkeep, both in terms of its built and social environment. City life, in other words, is much too diversely made by diverse people to be captured in the big plans on the drawing boards of a few professionals behind closed doors. Such big plans mostly stifle rather than reflect or enable the social life of actually existing cities and therefore should be scrapped and replaced by small plans emanating largely from city people themselves.

PLANNING THE POST-INDUSTRIAL, NEOLIBERAL CITY

By the late 1960s, city economies in the Global North were beginning to restructure away from manufacturing into services, as described in Chapters 4 and 5. At the same time, advances in transportation and communications were making it easier for businesses and people to locate just about anywhere. And, indeed, many took this opportunity to relocate, resulting in decreasing, and otherwise uncertain, tax revenue bases for most every former industrial city. All of this, combined with the rise of opposition to big plans and an emerging, newfound neoliberal faith in free markets in general meant that public planning on a city-wide scale was increasingly considered unviable and even undesirable. With dwindling tax revenue, such big city planning was also increasingly unaffordable.

City planning thus became more privatized, based on partnerships with private-sector actors and directed at increasing the tax revenue base more than anything else. As discussed also in Chapter 6, besides being a necessity due to the relocation of major businesses away from cities, the evolving neoliberal belief among city planners was, first, that partnerships with the private sector would lead to the more efficient provision of public services. Second, it was believed that decreasing public funds could be leveraged with private funds in this manner in order to maintain or even increase the level of public services available as a result of increasing tax revenue in the near future. So the emphasis in this neoliberal privatization of city planning was put on attracting new development above anything else.

The main tools of such privatized city planning have been principally the **urban development corporation (UDC)** and the designation of tax-subsidized enterprise zones (EZs). UDCs are financed by a combination of public and private capital for the purpose of acquiring land in the city for redevelopment. Such corporations are run largely on a private profit basis with very little government oversight or regulation. They are a prime example of leveraging public with private funds in order to have a larger impact upon the redevelopment process. Large-scale examples of city redevelopment that were begun and managed by UDCs include the massive remodeling of the London Docklands beginning in the 1980s, including the huge commercial center of Canary Wharf. Another major example of this type of privatized city planning is the redevelopment of Baltimore's inner harbor, which began in earnest in the 1970s and has since become the model of many more redeveloped waterfronts in the United States, from Boston to Seattle.

The other major tool of neoliberal planning, EZs, essentially are modeled on the free trade zones that now dot many a country in the Global South. Such free trade zones are slices of a country's territory set aside to attract foreign firms by, essentially, relinquishing sovereignty as well as by providing tax relief, or no tax whatsoever, and actual government subsidies for relocation. In other words, the attracting country agrees to allow foreign firms to operate according to their own rules and virtually tax free in such zones as a reward for bringing jobs to the country. City EZs operate largely on the same basis. Again, the idea is that once jobs are created in newly attracted firms in the EZ, and once some profit is made, future tax revenue in the city will be greater than that expended in tax relief and subsidy. Not surprisingly, this has not always been the case, as city after city provide the same incentives via the same policy mechanisms and with the same sales pitch.

In the end, both planning tools are part of a broader and now quite common understanding of city planning that focuses overwhelmingly on the attraction of firms and the more affluent gentry to cities above anything else. And, of course, as all cities offer the same sort of social and built amenities to this end, it makes this end itself even harder to attain, if at all.

FULLY PRIVATIZING THE GARDEN CITY

Neoliberal city planning is thereby more privatized and more market driven than earlier planning movements with the main goal being to renew city tax revenue. Some argue that this is not really planning at all, particularly given that the results have not generally trickled down to meet the needs of most city people outside the core gentrifying areas. Others argue that this is just a matter of success and time. In any case, it is a type of city planning that relies mostly on private-sector criteria, management, and market processes and, in this, it is a peculiar type of "public" planning indeed. Not surprisingly, other planning in the city has become even more privatized. This type of planning mostly takes place in suburban and exurban locations and involves quite comprehensive and large-scale master-planning of large-scale metropolitan communities. In addition to housing, such planned communities on the urban fringe increasingly include parks and recreational areas, conservation or "wild" areas, village centers and leisure domains, and even town centers, with boutique retail establishments and fine restaurants. These communities are privately developed for private profit and privately governed by homeowners' associations in order to retain most "tax" revenue for the benefit of the residents themselves in the form of homeowners' fees.

In many respects, these new suburban and exurban master-planned communities have realized the garden city ideal. They are usually green belted with plenty of internal green space, housing is adequate for all residents, automobility is usually restricted within them, rendering them pedestrian friendly, and they are governed locally. But, unlike the garden city ideal, all community amenities are furnished by private developers for private profit, housing is adequate only if one can afford the much higher prices demanded, and governance is only for the social welfare of those already with the means to buy into the community. As a result, these privatopian communities, as discussed in Chapter 7, are really socially segregated enclaves, usually even gated, for more fortunate city people to escape the social and environmental problems of the larger city even more fully than traditional suburban communities. That such gated communities have been planned and developed in virtually every country around the world at this point does not bode well for the future for the majority of city dwellers, in this respect.

And, yet, more recently this privatopian city planning has been taken even further. Associated with what has been called the "*new urban*" movement in planning, new master-planned metropolitan communities now include provisions for the attraction of business firms and the development of other city accoutrements such as actual retail/entertainment/banking downtowns, hospitals and other medical facilities, neighborhood schools, and even police and other emergency services. This new urban planning movement is essentially concerned with recreating the small town life of yesteryear by more densely packing housing and other built environments, avoiding automobility and attaining pedestrian friendliness throughout the community, limiting population, providing adequate green space, and constructing a total community where one could conceivably live, work, and play without ever leaving the gates.

New urban communities, in this way, sound like they are designed to be as radical a solution to city problems as that of the original garden city or the galactic metropolitan ideal of the RPAA. The new urban ideal also makes contemporary sense in that it seems to address issues of both social and environmental sustainability by providing diversity in housing style and employment opportunities at the same time as limiting automobility. One of the earliest and still best examples of such new urban townscapes is Celebration, Florida, which was built on former property of the Disney Company in the Orlando metropolitan area. Here, the built environment is designed to mimic small towns of the 1940s in the United States with a central focus on the "downtown" and walkability. Celebration is governed by a homeowners' association which enforces strict restrictions on how residents can behave and use their properties, from the kind of plants that can be planted in yards to the color of paint on the trim of their houses. Although not physically gated, Celebration is socially restricted by the fact that housing prices within it are significantly higher than those of surrounding areas.

Again, of most importance is that this new self-seclusion of the relatively wealthy is so complete that, conceivably, residents never have to leave such towns to experience or even merely observe the more complex, messy, and often much less fortunately situated city cultures outside their limits. This is surely the final step in the self-seclusion and privatization of wealthier city people. And, so, as such communities actually have materialized in places such as Laguna West (California), Kentlands (Maryland), and now worldwide in places such as Orchid Bay (Belize), Thimpu (Bhutan), and all over very rapidly urbanizing China, it is increasingly clear that new urban developments are not providing solutions to city problems as foretold precisely because they have catered largely to the needs of, and are populated overwhelmingly by, the most fortunate city dwellers. That new urban planning ideas are spreading rapidly and widely across the planet, then, should be cause for concern among those still hoping to solve the problems that affect entire city populations.

A SMARTER AND MORE SUSTAINABLE CITY FUTURE?

Like most city planning over its modern and postmodern history, new urban planning also suffers from a bias toward the design of the built environment as a means of solving social problems. Such environmental determinism largely ignores the most important economic, political, and even cultural causes of social and environmental problems in cities. No wonder, then, that the proposed solutions such planning comes up with usually fall far short of actually solving such problems. Yet, there is still much that can be learned from such city plans, from the original garden city to the contemporary new urbanist rendition. The concern for severely limiting the growth and actual size of cities, providing green spacing, thinking about alternative types of transportation, recreating small town life as nodes in larger metropolitan areas, providing diversity in housing to house, ideally, diversely situated city people, and other such ideals have, indeed, reemerged in the contemporary planning movements concerned with "smart growth" and "**urban sustainability**." The catchy titled smart growth movement really harkens back to at least the ideals of new towns in the sense of limiting city growth via actively enforced planning regulations and requiring green belt spacing between cities. It also includes focus on non-automobile-centered metropolitan transportation and a concern for more environmentally friendly economic and residential growth via a reduction in waste-generating consumption and the use of alternative energy sources such as wind and solar. While this specific movement arose in the United States, with Portland (Oregon) usually considered the poster child for the movement, its ideals have caught the eye of planners elsewhere.

The concern for urban sustainability, on the other hand, is a planning movement that is truly worldwide and has obvious connections to, and many of the design characteristics of, all the planning movements just mentioned. Yet, for the most part, those city planners concerned most with urban sustainability tend to be more radical in their city planning ideals in that they focus on cities in both the Global North and, significantly, the Global South. They also tend to focus on both environmental *and* social sustainability, the latter defined variously as some combination of economic, political, and even cultural security and welfare for all city dwellers, rich and poor. So urban sustainability is not just about limiting growth, providing green space and belts, and so on, in particular cities for particular people. Rather, urban sustainability is about attempting to ensure that all city people have decent lives and decent environmental conditions and not just in Northern cities but also in the now overpopulating Southern megacities where, importantly, most of the city population of the world will be living in the very near future. The urban sustainability movement thus asks difficult questions concerning environmental and social justice not only within individual cities but globally. It also is not afraid to demand critical reflection about how the world is economically, politically, culturally, and environmentally configured and why. Whether such questions are heard and much reflection takes place is, of course, an open question at this point. But, again, because the majority of people in the world now live in cities and an even greater majority will do so in the very near future, city planners really are dealing with problems that affect virtually everybody across the planet. If that is the case, then surely it is time again to think bold and big plans for city life on a global scale.

Box 9 Reorienting global urbanization: The resurgence of China's coastal cities

The twentieth century brought much internal turmoil in China, culminating in a Communist Revolution in 1949. The Communists were thoroughly biased against the coastal treaty ports, which they considered to be remnants of foreign imperialism and Chinese humiliation. A new Communist China was focused on self-defense and self-reliance and not on foreign-dominated overseas trade and investment. Economic development policy was aimed at creating cities of internal industrial "production" and not cities merely of trade and "consumption" like the treaty ports. In addition to internal production, Communist authorities emphasized a more balanced development between cities and the countryside by means of drastically slowing the pace of rural-to-urban migration, especially in coastal areas. Indeed, millions of city dwellers were actually deported to the countryside, particularly in the early years of Communist rule.

Specifically, a strict policy of citizen registration, the *hukou* system, was put in place which limited the movement of people from their birthplace. Essentially, those with rural *hukou* were restricted from movement and those with city *hukou* were able to take advantage of government welfare benefits that the Communists bestowed on city people. These benefits traditionally included much subsidized provision of housing and food, virtually guaranteed employment, and pensions and other income subsidies. Many argue that this restrictive *hukou* system as well as the deportation of city people after the Communist Revolution resulted in a long period of actual *under-urbanization* in China, given the size of its total population.

However, after a new group of Chinese Communist Party (CCP) members led by Deng Xiaoping attained power in China in 1978, a whole set of "reform" policies were instituted. Of importance for Chinese cities is that these reform policies entailed an opening up of the Chinese economy to more foreign trade and investment. Special Economic Zones (SEZs) were created in the cities of Shenzhen, Zhuhai, and Shantou in the southeastern coastal province of Guangzhou and in Xiamen in the nearby province of Fujian. SEZs were to be experimental areas of relatively open trade and inward foreign investment as a means to stimulate city development, and they were quite successful in this. Massive amounts of foreign investment flowed in to take advantage of tax breaks and very low-cost Chinese labor. This sparked a great spurt of rural-to-urban migration to these coastal cities and they grew rapidly in population and economic diversity as a result. Indeed, the SEZs were so successful in this way that 14 more coastal cities were opened up in 1984 and several more during the 1980s and 1990s, including the Pudong New District in the previously dominant treaty port of Shanghai in 1990.

Overall, the reforms of 1978 included not just trade policies, but also a relaxation of the *hukou* system, the formal decentralization of government authority, and a growing private market in city land rights (proxy ownership). These new policies have led, in turn, to the current situation in which China is urbanizing at a rate virtually unsurpassed in the world. And this hyper-urbanization of coastal China appears to be rendering Chinese city life more and more like that of other cities around the world. This includes rising housing costs and shortages of affordable housing, the increasing socio-spatial polarization of suburban wealth and inner-city poverty, and gleaming high-rise, luxurious central business districts (CBDs), and similar city issues discussed throughout this book. Yet the CCP in power maintains itself as "Communist" with all that this might mean in terms of a commitment to a more controlled, equitable form of city development. What the future holds for Chinese cities as a result of this rapid economic resurgence of coastal China thus depends upon how this new coastal patterning of city development is allowed to play itself out by Communist authorities.

FURTHER READING

A fine history of Western city planning can be found in Christine Boyer's *Dreaming the Rational City* (Cambridge, MA: MIT Press, 1983). For their part, Georg Glasze and others have edited a book, *Private Cities: Global and Local Perspectives* (London: Routledge, 2006), which explores the more recent privatization of planning internationally. A now classic examination of planning across a long history is Peter Hall's *Cities of Tomorrow: An Intellectual History of Urban Planning and Design* (Oxford, UK: Oxford University Press, 2001), whereas Jane Jacobs's *The Death and Life of Great American Cities* (New York, NY: Random House, 1961) provided a spark for a complete re-evaluation of the role of city planning in terms of the quality of city life in the face of mass urbanization in the post-World War II period. Floris Jan van Luyn provides a broad and readable contemporary account of city happenings in China in *A Floating City of Peasants: The Great Migration in Contemporary China* (New York: W. W. Norton, 2008). Finally, Steven Ward has edited a fine book of chapter accounts of various aspects of *The Garden City: Past, Present, and Future* (London: Routledge, 2011), which very much underscores a certain continuity of this idea in the history of Western city planning, in general.

CITY FUTURES

Since the majority of the world's population of over 7 billion people now live in cities, city issues, at this point, truly are global issues. What happens in cities and because of city life therefore should matter to all of us, whether one lives in the Global North or in the Global South. City life entails unique forms of human economic, political, and cultural relations which must be adapted to in order to ensure relative social stability and progress. Historically, the growth of cities brought much social change and no little turmoil in its wake, as many new city people could not be accommodated immediately or adequately with shelter, economic security, or personal safety. Because of this, many city people not only had much trouble adapting to their new, densely packed living arrangements, but they also soon grew quite restless with their situation. Indeed, the story of how this restlessness was eased, or not, over time is really the story of the creation of human civilization itself, as this book has argued in broad overview.

A TROUBLESOME CITY FUTURE?

So, what are the emerging social and environmental issues of today's newly citified and increasingly restless global civilization in the making? There are four highly linked and equally troubling trends that can be discerned. The first concerns the rate at which cities are growing in the Global South. It is in relatively poor post-colonial countries where cities are growing the fastest and this should be great cause for concern. As a result of public health breakthroughs leading to rapid population growth combined with few economic opportunities in rural areas, people are literally flocking to cities in countries with little capacity to provide for their new city-based public needs. The fact that such countries usually have only one or two primate cities as a result of Western imperialism means, in turn, that their cities have become quite over-urbanized megacities. As a result, it may not literally be a "planet of slums" as one recent commentator has put it, but it soon may be as literally billions of people try to eke out a simple living in the majority poor and underserviced districts of such enormous megacities.

Regardless of the issues of social justice that this situation in today's megacities of the Global South should elicit in one's mind, such dense concentrations of destitute people surely represent a ticking time bomb both in social and environmental terms. Even if the public finances were available, no public security force could hope to cover the whole city in order to keep the peace, which means that most city people will continually have to fend for themselves. Similarly, emergency services, from fire to medical, are unavailable to most city people in the ever-growing slums. Finally, with very little chance of being supplied with adequate water, sewage, and waste-disposal services, this social destitution is simply exacerbated by environmental degradation in major parts of most megacities today.

The material point is that such dire living conditions endured by most of the world's city people today will likely lead to much social unrest in the very near future. And such social unrest is not likely to be contained within city limits. Rather, it will spill out and engulf whole countries, if not larger regions. Arguably, the recent so-called "Arab Spring" can be seen as not just a new media-driven affair, in this regard, but very much a city-based one. Furthermore, slum-like living conditions in megacities of the Global South combined with the lack of opportunities in their immediate hinterlands is a major factor behind most individual decisions to migrate to the Global North, whether legally or illegally, in a desperate quest for mere survival. As a result, the problems associated with megacity over-urbanization already have come to be felt in cities in the North. Finally, because of the concentrated ecological effects of ever-growing numbers of city people, ever-wider swathes of nature are being seriously degraded around megacities which threaten to make such city life even more dire for most in addition to destroying major globally important ecosystem functions.

The second troubling trend in cities around the world are the effects of the notion, now quite common, that government is always less efficient than the private sector in terms of ensuring city development. In both the Global North and South, such a neoliberal persuasion has led to a situation where most cities have gleaming downtowns or sparkling waterfronts full of tourist and entertainment facilities and high-rent office space and housing, but much of the rest of the city is largely being ignored in the development process. As a result, no little polarization has come to exist in most cities around the world, both in built and social terms. While this polarization may not be as manifest as those now speaking about "dual cities" have it, it is clear that more and more cities are experiencing such polarization, largely due to the reliance on private sector-driven city development policies. In this, those who have the wherewithal to inhabit and work within such gleaming downtowns have more and more in common with their peers in other downtowns elsewhere than with their fellow city people in much closer proximity. And to the extent that this is the case, care for the living conditions of such fellow city dwellers, if it exists at all, is likely to wane on the part of city elite, thereby exacerbating an already polarized situation.

A third troubling trend in today's cities is simply reinforcing this growing dualism within them. This has to do with what can be called the Americanization of city and suburban communities across the globe. There is a growing automobile-based, high-consumption lifestyle that is rapidly circulating the globe and lighting down most particularly in cities. The environmental consequences of this lifestyle are considerable, particularly considering the current size of the ecological footprint the population of the United States itself now imprints upon global nature. Global carrying capacity simply cannot accommodate such pressure if all downtown city people across the planet were to adopt such a lifestyle. And surely this is cause for great concern.

Here, however, the emphasis is on the way in which city life is being consciously organized. City policy and planning is now dominated globally by planning movements that have resulted in the ever-greater spatial seclusion of more fortunate city people behind gates guarded by wealth and privilege, if not also by guards and steel rails. Gated anti-urban, privatopian communities of like life-styled and like-minded people are popping up even in the Global South and this surely does not bode well for entire city futures. This is especially the case to the extent that such gated communities now include their own downtowns in addition to leisure amenities as the new urbanists would have it. Again, conceivably, in this context city elites would never have to leave the gates of their privatopian communities to even see, let alone do anything about, the city problems around them.

A final aspect of a more citified world that should be of concern has to do with the relationship of city people to non-human nature. That city life has great impact upon non-human nature has been discussed in various ways throughout this book and especially in Chapter 9. That this impact is growing rapidly with continuing rapid urbanization trends in the early twenty-first century is of no doubt. Here, however, the emphasis is more speculative. City people have a much different day-to-day relationship with non-human nature than those leading a more rural-based lifestyle. This includes city people enclosed in mostly ersatz, countrified privatopian communities in more rural suburbs and exurbs. What non-human nature exists in cities and suburbs is always more humanized and otherwise orchestrated by human management than in more rural areas. City people, in this respect, are more fully alienated from real natural processes and phenomena than their rural counterparts. The fear is that the more the world's population consists of full-time city people, the less knowledge of, and then care for, non-human nature there is likely to be in the same manner that downtown and privatopian people largely have lost care for their fellow, less fortunate city dwellers. And this certainly does not bode well at all for the planet and, then, for our very survival as a species.

OR, RATHER, THE FUTURE PROMISE OF CITY-BASED GLOBALISM?

These appear to be troublesome trends, indeed. But the extent of their troublesomeness depends upon what exactly is made of them by city people themselves. This is the very promise of city life. Here, then, focus will be put on four essentially linked potential benefits of a global city future. First, cities are the very places where one has freedom to do things differently. In traditional rural communities, the existing *Gemeinschaft* social relations essentially are conformist, where expected social behavior is well known and well policed. One knows one's place, to be sure, and one needs to stay in one's place by performing in expected ways. This is the irony of those contemporary city planners who yearn to recreate such communities. Instead of solving real city problems in innovative ways, what is reconstructed is usually a built façade hiding an even thinner social conformity focused on property value alone. One must keep one's place in such cases according to a written housing code or else be forever exiled. In this, much city planning today is the ultimate in anti-city dogma.

Second, because real cities allow social freedom they are the sites of social innovation. By rubbing shoulders day to day with very different people in very close proximity, one cannot help but experience different, and in many cases much different, ways of doing things which can lead to much self-reflection. City people also see and hear different things and often must negotiate such difference every day. City people must, then, create their own culture(s) because whatever traditional culture they may have had once does not provide much guidance for coping with the social life of cities. To the extent that this is the case, then, it will be city people who come up with better solutions to the very social and environmental dualism of today's cities in which they dwell. City people have always been the source of innovations in human civilization in this way, as discussed throughout this book.

Third, because of this need to negotiate such social and cultural difference within them, cities are sites in which knowledge of a common humanity is more likely to arise than not. In other words, negotiating social and cultural difference necessitates finding some common ground to relate to others, even those whose lifestyle is quite different. The more, and more regularly, such negotiation takes place, then, the more this common human ground can be found among even very sharp differences and, in the best of circumstances, acted on in daily life. Indeed, that more and more people live in cities may mean that a truly **cosmopolitan** global culture can be constructed in the near future. This would be a culture based on this very negotiation of social and cultural difference that takes place in cities all the time, where people of different lifestyles will be able to assimilate *with* each other instead of *to* a particular, now mostly Western, lifestyle. In short, if such a global cosmopolitan ideal is to be realized at all, it will be only by the will of experienced city people.

Finally, it is clear to most thoughtful people that the human impact upon global non-human nature is not sustainable, at least at current rates of consumption and generation of waste. That these rates are likely to increase to the extent that a Western, if not a fully American, lifestyle becomes the norm on a global scale among the more wealthy should give everyone pause in this respect. Here it is important to note that those most concerned about cities have been most proactive in getting the message out about the need to adopt more sustainable living patterns, in general. Indeed, there is now a vast literature and no little actual practice concerned with the need to retrofit, or construct a new, city life, from economic relations to the built environment, in order to live more sustainably with nature. In short, it has been reflection on the very large and growing ecological footprint of cities around the planet that has forced this issue to the main table of human discussion and, hopefully, to ever-more innovative human action to confront it.

In the end, city life is what has made humans more distinctly human. That the long history of city development has also made many people more inhumane in their relations with their fellow human beings, as discussed throughout this book, should not divert attention from this point. Cities have been the source of essentially all the great innovations of human civilization, and now that city people make up the majority of the world's population it is certain that many more innovations will be forthcoming perhaps on the coattails of a truly cosmo*politan* global culture in the making. In the end, if the future global world is to consist of a planet of city slums and generally unjust social polarization and oppression for the majority of the world's population, it will not be the result of too many city people. Rather, it will be so because too many people have ignored or attempted to circumvent the very promise of social innovation and progress that city life necessarily entails.

Box 10 A cosmopolitan future?

Cities are the source of civilization because they are the places where social innovations emerge, whether economic, political, or cultural. City people have had more time to do other things besides toiling in the agricultural fields all day and these other things, from star gazing to philosophy to engineering and art to accounting and merchandizing to computer and information science and so on, are the very stuff of human civilization. Such social innovations are also the result of the very diversity of people packed, and continually mixing, together in tight spaces. In short, city people continue to be at the cutting edge of social innovation and such innovation generally spreads from city to countryside.

Now that the population of the globe includes an ever-growing majority of city people, it makes sense, then, to ponder what major social innovations, now quite global in scope, may be in store in the foreseeable future. Certainly, most recent events, from the Occupy Wall Street movement that has spread from city to city globally, to the city street protests over fiscal austerity measures which have arisen in many more cities in addition to Athens, to the Arab Spring still waiting to arrive in Damascus, Manama, and Sana'a, are types of city-based movements that are leading to social innovations even if the precise contours of such remain unclear. Here focus is on another trend that appears quite hopeful: the possible rise of a city-based, more cosmopolitan sensibility among more and more of the world's population.

People, goods, money, and information are moving much more rapidly across the planet, and cities are where most of this movement begins and ends. This has rendered cities much more diverse in population, economic characteristics, and informational access. Rather than a homogenizing Westernization of the world's cities, what appears to be happening on city streets is a vast and deep cultural mixing as salsa surpasses catsup in the United States, Bollywood and Nollywood influence Hollywood, rap and hip hop music is adopted and yet culturally modified by city youth around the world, and Al Jazeera competes with CNN to tell the world news. This cultural *mélange* is taking place, in turn, most intensely in cities. The most hopeful view is that as the world continues to urbanize, the more this city-sourced innovation of cultural cosmopolitanism will spread and then humankind will understand itself better to be made of one cloth, however multi-textured.

Yet, this may be simply a pipedream, particularly considering the many city-based social conflicts and continuing injustices, both social and environmental, that are brought to bear on some city people by other city people. Indeed, most talk of cosmopolitanism in today's world is coming from the mouths of the already global Westernized elite who do, indeed, live in similar luxurious built and social environments in today's major cities and who do travel the world rapidly and at will from Hilton Hotel (pick a name) to Hilton Hotel to Hilton Hotel. And yet … and yet the present narrative will close on a more optimistic note and that is this: there is a real possibility of a growing cosmopolitan city-based global culture, but only if the proper social circumstances are constructed as a result of social activism on the part of many more city people. The most important of these circumstances will be the increasing ability to conceptualize, and then to help establish in reality, a true cultural mixing where each diverse ingredient is treated in a relatively equal manner. This is a truly cosmopolitan ideal in which, as noted in the main text, city cultures of the future will be ones where each cultural ingredient assimilates *with* each other more or less equally instead of diverse cultural ingredients assimilating *to* some overwhelmingly dominant one, like that of the West. Such a hope may seem utopian, given the way in which the world has been structured economically, politically, and culturally to date, but if any concrete movement is to take place in this truly cosmopolitan direction it will surely be a city-based movement for all the reasons offered in this little book.

FURTHER READING

Rowland Atkinson and Gary Bridges provide a fine edited book on *Gentrification in a Global Context: The New Urban Colonialism* (London: Routledge, 2005), the merits of which are that it takes an international approach to this issue. Similarly, the edited volume by Neil Brenner and Nik Theodore, *The Global Cities Reader* (London: Routledge, 2006), treats cities both in the Global North and South. For his part, Robert Fine provides a good introduction to the topic of *Cosmopolitanism* (New York, NY: Routledge, 2007) at least in terms of politics. Josef Gugler's edited book *World Cities beyond the West: Globalization, Development, and Inequality* (Cambridge, UK:

Cambridge University Press, 2004) makes it clear that cities in the Global South are not just secondarily, or even lower, tiered megacities with no global power. For their part, Libby Porter and Kate Shaw have edited an important book concerning the impacts of contemporary city redevelopment schemes: *Whose Urban Renaissance? An International Comparison of Urban Regeneration Strategies* (New York, NY: Routledge, 2009). André Sørensen and Junichiro Okata remind us that over-urbanization in the Global South is an extremely important issue in their edited volume *Megacities: Urban Form, Governance, and Sustainability* (Tokyo: Springer, 2011), while Lee Trepanier and Khalil M. Habib have edited an important book concerning *Cosmopolitanism in an Age of Globalization: Citizens without States* (Lexington, MA: University Press of Kentucky, 2011), which very much muses on emerging cultural process of certain importance for city people both now and in the foreseeable future.

GLOSSARY

Acid precipitation: Any form of precipitation (rain, snow, fog, etc.) which mixes with chemical (particularly sulfuric and nitric) acids, which are emitted into the air usually from industrial processes in cities.

Agglomeration economies: These are economic benefits that a business accrues by making use of specialized supply and labor networks and resources that another business already has established in a particular location. Such economic benefits that so accrue to later locators is a major reason why similar businesses tend to locate in proximity to one another in the city.

Agricultural Revolution: Refers to the time, usually considered to be about 10,000 years ago, in which humankind began to settle down permanently in certain regions of the world to subsist on regularly farmed crops instead of hunting and gathering.

Albedo: Sometimes referred to as a "reflection coefficient," this is the power by which different surfaces reflect sunlight. Because cities consist of so much humanly constructed surface, they tend to reflect more sunlight (solar energy) back into the atmosphere, thereby affecting the local climate.

Assembly line: A process by which the production of a commodity is broken up into discrete and partial stages among several workers. Each worker thus completes only one stage in the entire production process – say, attaching windshields to automobiles – all day long. The development of such a production process was, and remains, important for cities because workers no longer need much skill to perform their duties. Unskilled labor thus becomes an important segment of the population of industrializing cities. Also, because of this need for unskilled labor, assembly-line firms look to (re)locate in places with the lowest possible wages.

Back office: Office jobs which demand few skills and little education, which are therefore mostly low wage such as those found in call centers and other customer service-type firms.

Bedroom communities: Places in the city, usually in the suburbs, where people live but must commute elsewhere to their jobs.

Bid-rent curve: What causes the rent value of property to generally rise toward the center of cities, rendering it necessary to build more vertically. There are many more people and firms "bidding" for such property than other locations in the city, thereby causing high rents.

Boosters: A term commonly used in the city literature to describe coalitions of city-based actors who actively promote their city as a good location to do business and in which to live.

Break of bulk sites: These are sites in which cargo is transferred from one type of transportation mode and network to another – say, from railroad to trucks and trucks to ships. Cities and towns generally arise on such sites.

BRICS: An acronym for **Br**azil, **I**ndia, **C**hina, and **S**outh Africa as the largest and fastest growing economies of the Global South.

Brownfield: A former industrial site in the city that is either abandoned or underused that may be developed again. Such sites generally are contaminated by toxic and therefore hazardous, or "brown," industrial effluents that linger from the former usage and must be cleaned up before any redevelopment can take place.

Carrying capacity: A way in which to talk about the natural limits of the Earth in terms of its capacity to support human consumption and waste.

Cash crops: These are agricultural crops that are produced solely for exchange in the marketplace and not necessarily for food for the producers. Large-scale cash cropping on the best lands was established in countries imperialized by Europe as a means to derive something of value for Europe. This pattern of production has continued for most post-colonial countries because of the difficulty, both in terms of economic disruption and financing, of reorienting production in a more internally rational way.

Central business district (CBD): Historically, the part of the city where the most important business firms and activities are located, usually in the most central areas of the city. In post-industrial metropolitan regions in the Global North and in megacities in the Global South, there are usually many CBDs in the sprawling suburbs and exurbs in a multi-nodal fashion.

Chain migration: The way in which members of the same rural village or ethnic group tend to migrate to cities by following the "chain" of earlier migrants and settling among their kind in specific areas of the city. This is done for safety in strange environments and other community self-help within the city.

City beautiful: A planning movement which emerged in the United States at the turn of the twentieth century that sought to render cities more aesthetically pleasing by cleaning up streets and constructing civically important and architecturally distinct public monuments and buildings, as well as open-access parks, in order to attract and maintain business.

City-states: These are cities the citizens of which have achieved formal political independence and therefore the right to conduct their political affairs on the basis of their own decisions, however way in which these ultimately are determined.

Collaborative planning: A planning movement which emerged during the 1960s which promotes the idea that city planning should not be a top-down process left solely to professional planners but, rather, a process which includes the participation of city people, from business interests to community activists to ordinary citizens, in the making of city plans.

Cosmopolitan: The ideal of a truly global culture formed as a new "polis" community on the basis of a negotiated and relatively equal assimilation of cultural differences.

Culture of poverty: The idea that environments of poverty in the city are elicited by, and at the same time elicit, poverty-stricken behavior among the poor as a result of the perceived lack of opportunity and general hopelessness engendered by their surroundings. Some consider this culture the main reason that many city people never escape poverty. Others suggest that this notion is merely a veil for blaming the victim and that poverty is much more the result of political/economic/social forces much beyond the control of the city poor.

Diseconomies of agglomeration: The added costs associated with overly congested city locations accruing from such things as added time to deliver goods and services due to traffic, higher rents due to popular central locations, and more intensive competition for skilled labor and higher salaries demanded.

Dual city: The idea that contemporary cities exhibit a highly polarized social morphology between the very rich and the very poor manifested in the contrast between gentrified CBDs and deteriorating inner cities. This is the result, it is argued, of the general neoliberal roll-back of the social welfare role of government and the increasing necessity of city policy-makers to fend for themselves with regard to economic development.

Ecological footprint: This is a specific estimate of how much nature is needed to meet the consumption and waste needs of a specific group of people, such as a city's population (or the entire population of the Earth), based on current patterns of consumption and waste generation.

Economies of scale: Economic benefits that accrue to larger scales of production due largely to standardization and bulk pricing of inputs.

Edge cities: A term coined by the American journalist Joel Garreau during the early 1990s in order to underscore the fact that the suburbs of large cities were beginning to sprout their own "downtowns" and CBDs with their own "suburbs," and so on, in a seemingly never-ending sprawl toward the exurbs.

Electronic waste: Also known as "e-waste," such waste is a growing problem in this increasingly digital and cell phone world. The chemical and other hazardous material in the hardware of communications and entertainment technology is difficult to dispose of and does not degrade rapidly. This is an increasingly toxic hazard in cities, particularly those of the Global South where more and more of this waste is dumped.

Elitist/pluralist regimes: Two types of city political regimes, the first dominated by public and private elite actors and the other more open to other, less elite, actors. Pluralist regimes thus are considered to be more democratic and inclusive in their policy-making, in general.

Eminent domain: The power of government to take private property for public use with what is considered just compensation extended to private property owners. In the United States, this power was particularly important for the construction of intra-urban highways and urban renewal policies.

Enterprise zones: Areas of the city that are demarcated by policy-makers expressly to attract business to locate there on the basis of favorable land prices, tax incentives, tax holidays, and direct public subsidy. This city planning tool is based on the model of free trade zones, which have been created in many countries of the Global South.

Environmental determinism: The idea that the environment directly determines human behavior. City planners and architects generally have such a bias in terms of believing that a certain type of built environment will elicit a certain type of human behavior.

Exurbs: Effectively the semi-rural suburbs of citifying suburbs which will, in turn, become more citified in a never-ending sprawl of the city.

Fiscal squeeze: The intense pressure put on public finance in cities which have lost much of their tax base due to the suburbanization of business and more wealthy people at the same time as they are faced with dwindling financial support from state, regional, and federal governments.

Fordism: A label given to the post-World War II period of development in the Global North in which there was a compromise among big government, big labor, and big business to ensure continually smooth economic growth via social welfare and infrastructure investment, the maintenance of relative peace in the workplace, and regular raises in wages and salaries.

Garden cities: Originally conceived by Ebenezer Howard (1850–1928), garden cities were to be cities of limited population surrounded by greenbelts and co-operatively owned and managed by their residents. This was a radical idea for a type of metropolitan regionalization based on networks of such cities connected by public transportation as an alternative to over-urbanizing, ever-sprawling industrial cities.

Gatekeepers of housing: Individuals who work in banks and real estate companies, as well as builders and property developers who have the power to either facilitate the acquisition of adequate housing or make it more difficult.

***Gemeinschaft*:** A German word that is usually translated as "community." Used by Ferdinand Tönnies, a German social historian, and other urban theorists to refer to the type of close-knit social relations of small village life where everyone knows everyone and community is held together by common social norms and regulations.

Gentrification: The process by which the gentry (higher-income groups) reinvest and thereby renew parts of the built environments of cities to their own tastes. This process generally crowds out the non-gentry from these areas due to higher rents and more restrictions on property usage, putting added pressure on the resources of the remaining low-rent areas of the city.

***Gesellschaft*:** A German word that is usually translated as "society." Used by Tönnies and others to refer to the type of loose-knit, temporary, and utilitarian social relations of cities where everyone is a relative stranger to everyone else and society is held together, or not, by impersonal, publicly negotiated norms and regulations.

Ghettoization: The process by which a group of city people is confined by active discrimination (if not actual force) to live in particular areas of the city.

Global city: A city, like London, Tokyo, and New York, that contains many headquarters of firms most active and powerful in the global economy. In this age of globalization and neoliberalism, it is often said that all cities need to aspire to this status by attracting ever-greater numbers of such headquarters and globalizing firms.

Global North: While not quite precise geographically (think Australia, New Zealand), this refers to relatively rich, increasingly post-industrial countries of the world.

Global South: Equally imprecise geographically, this refers to the relatively poorer, mostly post-colonial countries of the world.

Globalization: A now ubiquitous term that points to what many see as the economic, political, cultural, and environmental processes that have overflowed national state boundaries and regulations, thereby rendering the world's peoples more and more in constant, increasingly digital, contact with each other.

Governance: Governing the city on the basis of some combination of public and private authority. This notion describes the increasing use of public–private partnerships and the outsourcing of formerly public services to private firms, from economic development policies to transportation to schools, etc.

Green belt: A planned and enforced no-build zone surrounding the city that is meant to limit sprawl and to protect rural agricultural and natural areas.

Greenfield sites: Sites which have never been developed and therefore are generally cheaper to buy or rent and then build structures on.

Growth machine: Refers to the type of city-based policy that is singularly focused on economic development as opposed to more social welfare issues. Typically, such policy was derived from more elitist political regimes; but, more recently, it has become a policy of necessity as a result of the rise of neoliberalism.

Haussmannism: A term used to denote modern top-down, large-scale, monumental city planning in the mode that Georges-Eugène Haussmann undertook in Paris during the late 1800s.

Heat island: Because of the nature of the albedo of city surfaces, cities tend to create localized hot spots of reflected solar energy, thereby affecting the climate around them.

Hinterlands: Traditionally, this term connotes the rural area directly outside of a city's limits where raw material and human resources for city development are sourced. In today's globalizing world, however, a city's hinterland can be truly global both in terms of such resources but also in terms of its ecological impact.

Historical preservation: A planning movement aimed at protecting historically important and unique built environments of cities usually via national designation, and restricted modification and usage, of private property within specific districts.

Imagineered: A term borrowed from the Disney Company which points to the way in which cities and residential areas are imaginatively packaged as prime locations for business and living on the basis of the avoidance, or active smothering, of social and built difference and problems derived thereof under a façade of happy safety and sameness.

Industrial restructuring: Usually refers to the process by which an individual firm restructures its production process so that its headquarters can be located in a place other than that of its assembly plant or that of its customer service center. It also refers to the manner in which the economies of the Global North have become predominantly service-sector based while those of the economies of the Global South are becoming more and more industrial.

Industrial Revolution: A period of Western history (ca. 1750) marked by the move from predominantly rural agricultural production and living to city-based industrialization. This is the period in which urbanization as the continuous process of rural-to-urban migration begins.

Informal sector: The part of the city economy that is off the government books, so to speak. This includes jobs which either are illegal, such as drug dealing, prostitution, and garbage picking, or are being undertaken secretly, such as sweat shops and other illegal manufacturing.

Inner city: Traditionally the residential areas in the center of the city surrounding the CBD, including industrial districts. Today it connotes the more poverty-striken parts of this area as other areas have been gentrified of late.

International division of labor: A situation in which regions of the world specialize in certain kinds of production and then trade with each other. The global reality is that countries in the Global South are specialized in the production of raw and semi-processed goods and those of the Global North in industrial and post-industrial goods and services as a legacy of European imperialism. Some argue that such a division of labor is a form of neocolonialism from which poorer countries can never hope to be free.

Labor intensive: Production processes that require many laborers to accomplish.

Leachate: Any fluid that passes through matter, picking up particulates and/or dissolving environmentally hazardous materials from such matter. It is a particularly harmful effluent from dumps and landfills.

Mass production and mass consumption: These phenomena involve the spread of assembly-line industrial production techniques throughout industrializing countries. More specifically, mass production involves the standardization of such production techniques, as well as products, on the basis of interchangeable, unskilled workers and commodity parts. Such standardization resulted in such economies in production that the prices of most commodities dropped and increasing numbers of people could afford them. Mass consumption was also enhanced by conscious policy on the part of Henry Ford and his imitators to raise wages and to offer installment credit, as well as by government welfare policies.

McMansion: An oversized, relatively expensive house which is usually built with much style and flair to connote unique sophistication but most often simply mimics other such houses down the street as a result of standardized mass production building techniques.

Medical revolution: The third demographic revolution in human history, along with the agricultural and industrial revolutions. This began in the post-World War II period and involved the spread of medical inventions and innovations in public health around the world, including to post-colonial countries. Such a spread of modern medical innovations has led to a dramatic decrease in infant mortality, particularly in the Global South.

Megacity: An enormous primate city of the Global South that suffers from severe over-urbanization with small islands of gentrified bits in a relative sea of slums.

Mode and network of transportation: The mode of transportation is the actual means of conveyance, such as horse and buggy, rail cars, or automobiles, and the network entails the nature of the routes the mode needs to get around, such as trails, rail lines, and roads. Changes over time in both aspects of transportation are important to consider as such changes have had enormous impact upon the development of cities over time.

Monumentalism: The construction of relatively massive built structures in the city to reflect social and economic power or otherwise adorn the city in grandly built fashion, including wide boulevards and parks.

Morphology: This is the spatial layout of the city, both in terms of buildings and people. In this book, this term is interchanged with that of environment, as in the built and social "environment."

Multiplier effect: The way in which wages spent on other things than what is actually produced creates more wages spent on still other things, and so on and so on.

Neocolonial: The maintenance of dependency relations between the Global North and the Global South even after formal imperialism by way of vast differentials in political and economic power in the world. In this book, focus is on the creation and now maintenance of production patterns in which countries of the Global South are locked into raw materials production or as permanent sources of relatively unskilled and thus low-wage labor.

Neoliberalism: An increasingly hegemonic political ideology which suggests that the private sector and free markets are always more efficient at providing for economic growth and social well-being than government. Therefore, politicians should free markets as much as possible from taxes and regulation, and government itself should be scaled back dramatically and privatized as much as possible in order to be run in a more business-like manner.

Nested hierarchy of cities: This is a city structure within a country in which there are very large cities, somewhat large cities, somewhat small cities, small cities, and then villages in a nice hierarchical pattern based on population size.

New town: Conceived as a fully master-planned alternative to urban sprawl in post-World War II England, new towns were designed to limit population in fully functioning small cities with green belts as a means to disperse city populations more evenly. Planned new towns can now be found in most every country.

New urban(ism): A very recent planning movement that arose in the United States which calls for the fully planned (re)creation of small towns, with housing, recreation, and jobs included. New urbanists believe that such towns will be an alternative to the automobility of contemporary cities and therefore more environmentally sustainable. To date, however, new urban communities have catered only to the most wealthy city dwellers.

Off-shore/off-shoring: The process where large corporations, usually with headquarters in the Global North, can locate the most labor-intensive parts of their production process to countries of the Global South with much lower-cost labor.

Outsourcing: When a business contracts out to other business functions such as management consulting, legal service, accounting, and maintenance and janitorial services, which could be, and often times traditionally were, performed by the business itself.

Over-urbanization: This is a situation in which a city's population has increased to a size much too big to be supplied with adequate water or for waste to be disposed of properly or for personal security to be assured or for adequate housing to be supplied, etc.

Patriarchy: The social construction of gender relations in which males have much more social power than females.

Political regime: The conception that city policy is really formulated and driven by a coalition of public and private actors and their interests and not just by official politicians and public bureaucrats.

Post-industrial cities: Cities in which those employed in the formal economy are overwhelmingly in the service sector. This is increasingly the norm in cities of the Global North and in the gentrified bits of megacities in the Global South.

Primate city: A city the population of which is so much larger than the next-sized city that its overwhelming primacy is obvious. Most post-colonial countries have this sort of city structure in which one or two cities are much bigger than the next-sized city. This has led to severe problems of over-urbanization and the rise of megacities.

Privatopia: A term that refers to the growing trend toward the private governance of metropolitan communities in terms of the growing numbers of homeowners' associations or developer-ruled areas, many of which are physically set off from the rest of the metropolitan area by privately policed gates.

Public–private partnerships: Contractual arrangements between city governments and private-sector actors enabling the latter to provide public goods and services. Such partnerships are an example of the general contemporary trend toward the neoliberal privatization of city governance.

Redlining: A lending practice, now illegal, in which banks and other lending agencies would designate some parts of the city as absolute "no loan" areas due to the perceived high risk of default. In reality, such perception was tainted by racism and classism as much as strict economic criteria. It is called redlining because, before the practice was illegal, red markers were used on city maps to clearly identify such "no loan" areas.

Reformation: A movement (ca. 1500) within the Renaissance period initially led by Martin Luther aimed at reforming the Catholic Church. When this did not happen it became a protest movement resulting in a complete split within the Western Christian world between Catholics and Protestants. Protestant teaching was very important for the continuing rise of the city-based merchant class of Europe.

Renaissance: Literally "rebirth," as in the rebirth of city life and therefore civilization at the end of the Western Middle Ages (ca. 1250).

Skills mismatch: A situation in which the skills necessary for jobs in the CBD of cities no longer match those which most inner-city residents have, and the jobs that do match these latter are located in places these residents cannot effectively reach due to lack of adequate transportation.

Slum: An area of the city that is characterized by an inadequate, severely derelict built environment populated by the very poorest city people, mostly working, if at all, in the informal sector of the economy.

Smart growth: A recent planning movement which seeks to limit sprawl by enforcing territorial limits to the built environment and automobility. The poster child for this movement is Portland, Oregon.

Smog: Originally coined to denote the combination of factory **sm**oke and f**og** in industrializing London. Now the term is used more commonly to signify the dirty haze that lingers in the air over most cities.

Social capital: Generally refers to the level of social participation and cohesiveness of groups of individuals with higher levels being considered important for the social stability and future success of cities as more socially cohesive communities. There is no little controversy over this notion, particularly in terms of how it is defined as well as whether it can actually be measured. Some suggest that it is simply the contemporary rendition of traditional *Gemeinschaft*, with all that this means in terms of what it may connote.

Social division of labor: The pattern of labor in which some groups of people specialize in production in, say, agriculture, or particular agricultural pursuits, and others specialize in industry, or particular industrial pursuits. Such specialization thus necessitates trade for all necessities to be provided to each specialized group.

Streetcar suburbs: Literally those residential suburbs that arose along or at the end of streetcar tracks in a star-like fashion in early industrial cities.

Suburban infill: After industrialism, suburbs generally grew out from the city along fixed transportation networks such as streetcar lines and railways. This lent a generalized star-like pattern to the built environment. As transportation networks grew more dense and network-like, particularly with the rise of automobiles and ever-expanding miles of roadway, the areas between the points of the star began to fill with housing and other built structures.

Sunbelt: Regions of the U.S. South and Southwest which did not urbanize or industrialize until very recently and which thus had many greenfield sites and much low-wage labor to draw on for firms looking for lower-cost locations for production after the economic crisis of the 1970s.

Taylorism: General term which refers to intense time management studies carried out to determine the most efficient time in which a certain work task should be completed by a worker. Such studies, pioneered by Frederick Taylor early in the twentieth century, helped Henry Ford and imitators to determine how to pace the newly automated assembly-line production process.

Transit captivity: The fact that some city people are literally trapped in certain areas of the city as a result of lack of adequate personal transportation modes, as well as the very uneven coverage of available public transportation across the entire metropolitan region. This has become a particular problem in the United States as increasingly privatized suburban communities literally prevent public transport access.

Underclass: A term used to refer to those city people who are not employed in the formal economy and therefore not a part of the formal working class. Sometimes used in a pejorative manner, this term nevertheless underscores the lack of adequate job opportunities in many inner-city areas and the resulting loss of hope for economic success on the part of many inner-city residents.

Urban development corporation (UDC): A corporation that is funded both publicly and privately but acts like a private corporation/developer of city property to facilitate urban redevelopment.

Urban microclimates: The capacity of cities to create climatic conditions that are specific to them, particularly by modifying surface albedo and wind patterns.

Urban renewal: A set of policies aimed at clearing slums on the basis of land acquisition and redevelopment via the use of eminent domain. Combined with highway building, such federally funded land acquisition cleared much urban land in the United States, leaving it to city authorities to decide how best to redevelop this land. This often led to the displacement of large contingents of lower-income city residents, thereby putting demand pressure on ever-scarcer housing in other areas of the inner city.

Urban sprawl: The continual expansion of the built and social environment of the city ever farther out into the rural hinterland.

Urban sustainability: The idea that cities should be (re)designed to ensure environmental and, importantly, social sustainability in the sense of limiting their ecological footprint and ensuring social welfare for all.

Urbanization economies: Economic benefits that accrue to firms simply because of their location in large cities. These benefits include such things as access to a large skilled and diverse labor force, many specialized professional service firms, better educational resources for their employees, and a range of other city amenities such as libraries and public archives not available in smaller towns.

Vacancy chain: A process by which as families move up the income ladder they move to larger housing units out toward the suburbs. This, in turn, provides a vacancy in their former house for those families one step behind on the income ladder to fill.

Vertical integration: A process by which a firm attempts to capture all phases of production and distribution of its product, from the raw materials to the final commodity, under the firm's control. This was considered good business practice to ensure economies of scale and quality until very recently when firms have found "**vertical disintegration**" via outsourcing many of these phases of production to be more efficient in the post-Fordist competitive context.

Western: Refers to European and European-derived countries (particularly the United States, Australia, and New Zealand) of the world.

White collar worker: Workers who work in management and other high-skilled professional positions who take up most of the office jobs in post-industrial CBDs.

INDEX

Significant help compiling this index was provided by Jason Simms, M.A., M.P.H., CPH, Information Technology Support Specialist, at the University of South Florida.